d - Block Chemistry

Mark J. Winter

Department of Chemistry
The University of Sheffield

OXFORD
UNIVERSITY PRESS
Great Clarendon Street, Oxford OX2 6DP
Oxford University Press is a department of the University of Oxford.
It furthers the University's objective of excellence in research, scholarship,
and education by publishing worldwide in
Oxford New York
Auckland Cape Town Dar es Salaam Hong Kong Karachi
Kuala Lumpur Madrid Melbourne Mexico City Nairobi
New Delhi Shanghai Taipei Toronto
With offices in
Argentina Austria Brazil Chile Czech Republic France Greece
Guatemala Hungary Italy Japan South Korea Poland Portugal
Singapore Switzerland Thailand Turkey Ukraine Vietnam
Oxford is a registered trade mark of Oxford University Press
in the UK and in certain other countries
Published in the United States
by Oxford University Press Inc., New York

First published 1994
Reprinted 1996, 1998, 1999, 2000, 2001, 2003, 2004

A catalogue record for this book is available from the British Library
Library of Congress Cataloging in Publication Data
(Data available)
ISBN 0 19 855696 9
11
Printed and bound in Great Britain by
Antony Rowe Ltd, Chippenham, Wiltshire

Series Editor's Foreword

Transition metal chemistry is one of the most colourful aspects of inorganic chemistry, and an understanding of its basics is essential to a graduating chemist; it is also a component of sister sciences such as geology and biochemistry. The variety of physical and chemical behaviour is so wide that it is only by grasping the fundamentals that this diversity can be seen to be beautiful rather than perplexing.

Oxford Chemistry Primers are designed to give a concise introduction to all chemistry students by providing material that would usually be covered in an 8 to 10 lecture course. As well as giving up-to-date information, this series provides the explanations and rationales that are the framework of inorganic chemistry. Building on the background provided by his earlier Primer ('Chemical Bonding' – OCP 15), Mark Winter gives the reader an elegant presentation of the basis of current thinking of *d*-element chemistry, a component of all undergraduate chemistry courses.

John Evans
Department of Chemistry,
University of Southampton

Preface

This short book is intended to introduce some concepts of *d*-block chemistry in a clear and concise fashion. The material could form the basis of an introductory course in *d*-block metal chemistry. When commencing a chemistry course, many students find the array of structural types displayed by *d*-block metal complexes somewhat bewildering. It is hoped that the reader will come to appreciate that there is some order amongst the apparent chaos. The coverage is aimed at the student rather than the lecturer. It is anticipated that this short text could find a place *alongside* textbooks containing more detailed coverage.

I am very pleased to acknowledge the help of the Cambridge Crystallographic Data Centre in the construction of many of the molecular geometry pictures. The Cambridge Structural Database System is outlined in: F. H. Allen, J. E. Davies, J. J. Galloy, O. Johnson, O. Kennard, C. F. Macrae, E. M. Mitchell, G. F. Mitchell, J. M. Smith, and D. G. Watson, *Journal of Chemical Information and Computer Sciences*, 1991, 31, 187. The illustration on page 73 is of a stamp from the private collection of Professor C. M. Lang photographed by Mr. G. J. Shulfer, both of the University of Wisconsin – Stevens Point, USA.

Many people made constructive criticism during the preparation of this book, particularly David Fenton and Colin White, and I am very grateful to them for taking the time to do so. All the remaining errors and misconceptions are, of course, mine.

Mark Winter

Sheffield
May, 1994

Contents

1 Introduction

1.1 The *d*-block and transition elements

D.I. Mendeleev: 1834–1907

The term 'transition element' or 'transition metal' appears to derive from early studies of periodicity such as those epitomized by the Mendeleev periodic table (Table 1.1) of the elements. His horizontal table of the elements (1871) was an attempt to group the elements together in such a fashion that the chemistry of the elements might be rationalized and predicted. In this table there are eight groups labelled I–VIII, with each subdivided into A and B subgroups. Mendeleev recognized that certain properties of elements in Group VIII are related to those of some of the elements in Group VII and those at the start of the next row in Group I. In that sense, these elements might be described as possessing properties *transitional* from one row of the table to the next.

These days, the transition elements are often defined as those which as elements have valence *d* and/or *f* electrons. There seems to be a subtle difference in the definition of 'transition elements' taken from various sources. Some define the transition elements as those which as elements have partly filled *d* or *f* shells. Others use a broader definition to include those

Table 1.1. The Periodic Table in the Mendeleev format.

	I		II		III		IV		V		VI		VII		VIII
							RH_4		RH_3		RH_2		RH		
	R_2O		RO		R_2O_3		RO_2		R_2O_2		RO_3		R_2O_3		RO_4
Row	A	B	A	B	A	B	A	B	A	B	A	B	A	B	
1		1 H													
2	7 Li		9.4 Be		11 B		12 C		14 N		16 O		19 F		
3		23 Na		24 Mg		27.3 Al		28 Si		31 P		32 S		35.5 Cl	
4	39 K		40 Ca		44 –		48 Ti		51 V		52 Cr		55 Mn		56 Fe, 59 Co, 59 Ni, 63 Cu
5		(63 Cu)		65 Zn		68 –		72 –		75 As		78 Se		80 Br	
6	85 Rb		87 Sr		88 Yt?		90 Zr		94 Nb		96 Mo		100 –		104 Ru, 104 Rh, 106 Pd, 108 Ag
7		(108 Ag)		112 Cd		113 In		118 Sn		122 Sb		125 Te		127 I	
8	133 Cs		137 Ba		138 Di?		140 Ce?		–		–		–		– – – –
9		(–)		–		–		–		–		–		–	
10	–		–		178 Er?		180 La?		182 Ta		184 W		–		195 Os, 197 Ir, 198 Pt, 199 Au
11		(199 Au)		200 Hg		204 Tl		207 Pb		208 Bi		–		–	
12	–		–		–		231 Th		–		240 U		–		– – – –

elements that have partly filled *d* or *f* shells in their most common compounds. Compounds of Cu(II) ($3d^9$), Ag(II), ($4d^9$), and Au(III), ($5d^8$) are all known and so this second definition seems reasonable since it allows copper, silver, and gold (all of which display chemistry related to the other transition metal compounds) to be referred to as transition elements. Under these definitions, zinc, cadmium, and mercury are not transition elements.

The term *d-block elements* is unambiguous and refers to the thirty elements contained in the 10 columns (3 – 12) in the periodic table (inside back cover). It is also a convenient term since it includes the elements Zn, Cd, and Hg, some properties of which it is logically appropriate to include in a discussion of transition metal chemistry.

For pragmatic reasons of space, this book does not address transition elements with partly filled *f* orbitals (*f*-block elements) and it is therefore convenient to describe the contents of this text as addressing *d*-block chemistry. Since all the *d*-block elements are metallic, the term *d*-block *metals* is synonymous. The chemistry of the *d*-block metal compounds is an extremely important part of chemistry and is to be found in discussions of such diverse areas as analytical chemistry, organic synthesis, catalysis, and metal extraction from ores.

1.2 Occurrence of *d*-block metals

The transition elements are widely distributed throughout the Earth's crust and the oceans. The concentrations of iron in the crust are far greater than those of the other transition metals summed together but the distributions are more even in sea water. Concentrations of the first row *d*-block elements are all higher in the human body than in ocean water. Here also the iron concentration is high reflecting its biochemical importance to mammalian life. Molybdenum is unique amongst the second and third transition series in that it is a requirement for life.

1.3 *d*-Block metals in nature

The *d*-block metals are extremely important in nature. Most of the first row *d*-block elements are biologically necessary trace elements. The *d*-block metals are fundamental constituents of minerals. It is not the intention to discuss biological or mineralogical chemistry here but it is as well to be aware of a few instances of their occurrence.

Table 1.2. Abundances of the first row transition elements.

	Sc	Ti	V	Cr	Mn	Fe	Co	Ni	Cu	Zn
crust	16	5600	160	100	950	41000	20	80	50	75
ocean	6.1×10^{-7}	0.00048	0.0011	0.00018	0.00011	0.0001	< 0.00001	0.0001	0.00008	0.00005
human	0.0008	0.054	<0.0002	0.006 - 0.11	0.0016 - 0.075	447	0.0002 - 0.04	0.01 - 0.05	1.01	7.0

Crustal abundances given in p.p.m., ocean abundances given in p.p.m. for Atlantic ocean surface levels, human abundances given in mg dm^{-3} for blood. Source: Emsley, J. (1991) *The Elements*, (2nd edn.). Oxford University Press, Oxford, England.

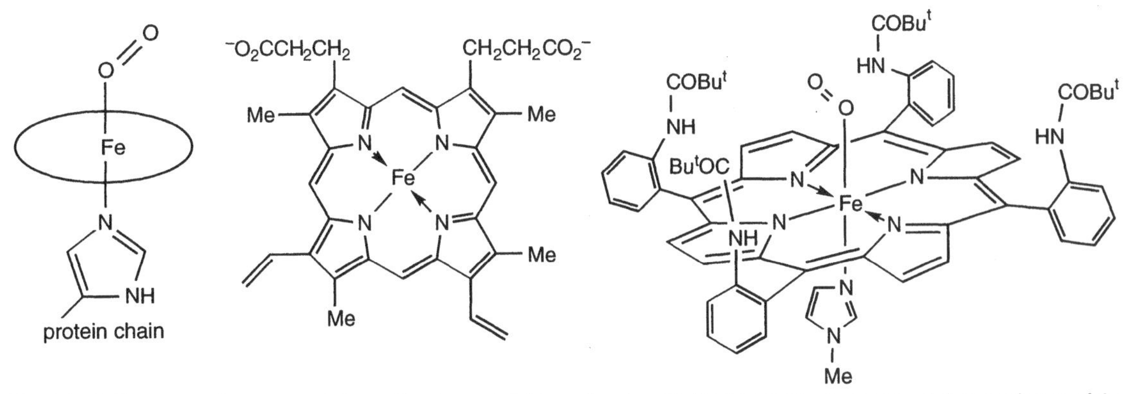

Fig. 1.1. A diagrammatic representation of the haemoglobin molecule, left, the haem group, centre, and an analogue of the haemoglobin molecule, right.

Oxygen transport and haemoglobin

Oxygen is transported around the bodies of mammals on haemoglobin (Fig 1.1), a large biological molecule. This molecule consists of iron bound inside a haem group and further attached to a complex protein chain. The whole is called haemoglobin.

Oxygen binds to iron [Fe(II)] under conditions of high oxygen partial pressure, that is, in the lungs. Haemoglobin transports oxygen to various tissues at which point the oxygen is released to another large biological molecule, myoglobin. One reasonable approach to an understanding of the chemistry of iron in the complex haemoglobin is to synthesize simpler analogues such as that shown in Fig. 1.1. By studying simpler model compounds such as this, perhaps the behaviour of the more complex biological molecule will become clearer.

Oxygen, O_2 (*dioxygen* to the inorganic chemist) uses a lone pair to bond to Fe(II) in the haemoglobin molecule. However other lone pair donors compete very effectively with O_2 for the iron electron pair acceptor site, particularly CO. Carbon monoxide is lethal for this reason: it prevents oxygen from being carried around the body.

Vitamin B_{12}

Vitamin B_{12}, cyanocobalamin, (Fig. 1.2) is another biologically important molecule. It is a compound of Co(III), but, again, one with a somewhat complex organic structure. Vitamin B_{12} is the only metal-containing vitamin and all higher mammals require it.

An important derivative of vitamin B_{12} is methylcobalamin; it has a similar structure but with methyl in place of the cyano group. It can transfer methyl groups to metals such as Hg(II), Tl(III), Pt(II), and Au(I). Methylcobalamin is an *organometallic* compound and is involved in the metabolism of methane-producing bacteria. Such bacteria are probably responsible for transforming elemental mercury into extremely toxic methylmercury species. One way to study vitamin B_{12} chemistry is to study much simpler and therefore more tractable analogues such as that shown in Fig 1.2.

Organometallic compounds are those that contain a metal bonded to an organic carbon group

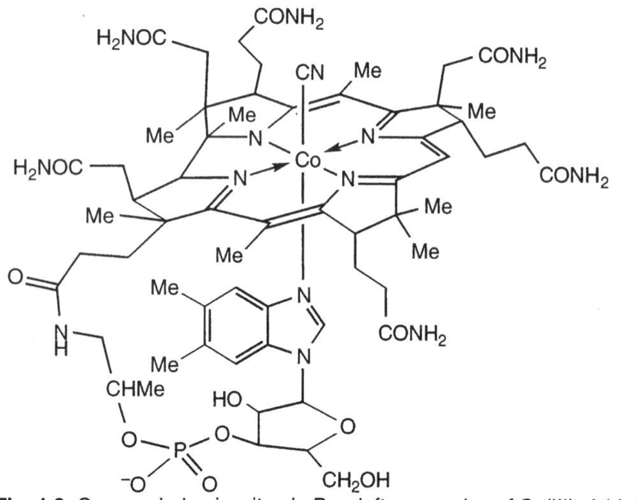

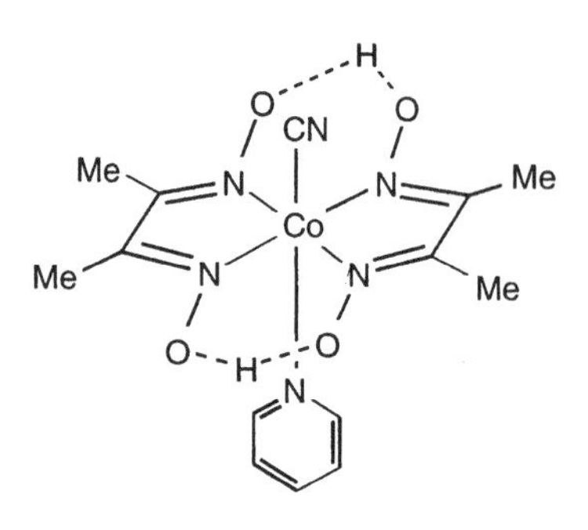

Fig. 1.2. Cyanocobalamin, vitamin B_{12}, left, a complex of Co(III). A bis(dimethylglyoximato)cobalt complex, or cobaloxime (right), is a model for vitamin B_{12}.

Enzyme: a biological catalyst

Nitrogen fixation

Nitrogen, N_2, is often referred to as *dinitrogen* by the coordination chemist. Each nitrogen atom has a lone pair which, in principle, is capable of bonding to a metal, although more often than not only one actually does so. It is this coordination which allows iron and molybdenum based *enzymes* in nitrogen 'fixing' bacteria to commence their work on N_2 gas to convert it into ammonia. Nitrogen fixing bacteria are found in the root nodules of plants such as clover and exist in a symbiotic relationship with the plant. It is the ability of these microbes to 'fix' nitrogen that makes clover such a valuable plant in crop rotation farming.

1.4 Gemstones

Although gemstones are expensive, they are often based on simple and common compounds. Ruby is based on *corundum*, a form of alumina, Al_2O_3, in which a few of the Al^{3+} ions are replaced by Cr^{3+} ion 'substituents'. In effect, this is a dilute *solid solution* of Cr_2O_3 in Al_2O_3. In ruby, the chromium ion is surrounded by six oxygen ions in a distorted octahedral array. The Cr—O bond distances are a little shorter than in the aquated chromium compound ion: $[Cr(OH_2)_6]^{3+}$.

If the contaminant in the alumina is Fe^{2+} and Ti^{4+} rather than Cr^{3+}, then the gemstone blue sapphire is produced instead. Other colours of sapphire gemstones are common. This is a consequence of differences in the *d*-block metal ion contaminants. It is therefore clear that the colours are a consequence of the transition metal substituent, rather than some bulk property of the Al_2O_3 skeleton.

Emeralds are based on a different mineral, the principal component of which is *beryl*, a beryllium-containing aluminosilicate. Some of the host lattice ions are again substituted by Cr(III), but in this case, the effect is to produce a green colour. Here also the Cr(III) geometry is distorted octahedral, but in this case the Cr—O bond distances are a little *longer* than in the $[Cr(OH_2)_6]^{3+}$ ion. The structure of beryl is more complex than the structure of corundum. It is possible to make large synthetic rubies, but it is apparently not yet possible to make synthetic emeralds.

The fact that it is rather easy to make rubies is of some importance since certain of their optical properties make them a useful material for lasers. The properties of such lasers hinge on the colours provided by the *d*-block metal ion substituents in the structure and so on the particular geometry displayed by Cr(III) in the ruby structure. Lasers can produce very intense sources of coherent radiation. This property has led to their use for the surgical attachment of retinas, the precision welding of high melting point metals, the measurement of the distance of the moon to the earth within an accuracy of centimetres, the forecasting of earthquakes through the precise measurements of movement across fault lines, artillery ranging, and in telecommunications where a modulated laser signal can carry many independent signals. Note that the *f*-block elements also make very good lasers which are in some ways more useful than lasers based on *d*-block elements. Such lasers are used in research on nuclear fusion where the intense light pulse is used to compress the fusible material and to heat it to 10^6–10^7K.

1.5 Diverse experimental observations

The chemistry of the *d*-block elements is one of the most diverse areas of chemistry, yet some of the most important fundamental chemical questions concerning *d*-block metals can be posed after executing a few extremely simple experiments. If one can begin to offer explanations for the results of these simple experiments, then that will be an impressive start towards developing an understanding of one of the most diverse areas of this fascinating area of science.

Drop a few iron filings into a little dilute sulphuric acid: they dissolve to form a very pale green coloured solution. Hydrogen gas is evolved. The corresponding reaction with manganese powder gives a faintly pink solution. The products of these two reactions are the iron(II) and the manganese(II) ions respectively. Copper metal also dissolves in sulphuric acid, in this case the product is the copper(II) ion, which in aqueous solution is a rather pale greenish blue colour. Add an aqueous ammonia solution to the copper(II) solution and the result is a deep genuine blue colour. Add potassium cyanide (KCN) to Fe(II) ion or Fe(III) ions: the compounds which can be isolated from these solutions have the empirical formula $K_4Fe(CN)_6$ and $K_3Fe(CN)_6$ respectively. Place a sample of each of these compounds in a sensitive balance and weigh it. Bring a strong magnet up to the sample and observe that the recorded weight changes, but in one case it *increases* and in the other it *decreases*.

iron(II), Fe(II): ferrous

copper(II), Cu(II): cupric

iron(III), Fe(III): ferric

Solutions containing the Fe(II) ion tend to oxidize to the Fe(III) ion. Add a little sodium thiocyanate, NaSCN, to solutions containing even a little iron(III) and the result is a remarkably intense blood red colour. This reaction is a highly sensitive test for the presence of Fe(III).

Pass a stream of carbon monoxide gas over nickel powder and the powder disappears! Place a condenser at the end of the reactor and a colourless volatile liquid is trapped as the nickel metal disappears. This liquid is nickel tetracarbonyl, $[Ni(CO)_4]$. A similar thing happens with iron but more forcing conditions are required. In this case the oily product is iron pentacarbonyl, $[Fe(CO)_5]$. These two carbonyl compounds are completely immiscible with water and their physical properties are very different from those of salts such as $K_4Fe(CN)_6$.

These are all simple experiments and mostly easy to carry out in any laboratory. But why are the various colours produced on dissolution of the metals in acid? Why is the colour produced in the SCN^- reaction so intense? Why are so many transition metal compounds coloured while most main group compounds are not? What is the nature of the hydrophobic metal carbonyl compounds? Why do the weights of the samples of $K_4Fe(CN)_6$ and $K_3Fe(CN)_6$ change in a magnetic field - *and in opposite fashions, one becoming lighter and the other heavier*?

The intention of this text is to develop an understanding of the properties of *d*-block metal compounds that can account for their colours (that is UV–visible spectroscopic properties) and magnetic properties. This understanding may be based on relatively simple ideas and be applicable to other properties of metal compounds. Remarkably, it is possible to use relatively simple ideas to rationalize the disparate experimental observations described above.

Alfred Werner: Nobel Laureate 1913

Alfred Werner and his colleagues did much to contribute to the understanding of *d*-block metal chemistry. An enormous body of experimental work and interpretation by Werner and his fellow workers resulted in his award of the Nobel prize in 1913. His ideas and modifications of these ideas are central to modern inorganic chemistry. Werner's training was in organic stereochemistry, but it was the concepts of that area of chemistry which were vital for the development of this area of inorganic chemistry.

1.6 Exercises

1. Review the definitions of 'transition metal' in your other text books.
2. Use your other text books to identify compounds found in biological systems other than those mentioned in this chapter which are *d*-block metal compounds.
3. For each of the *d*-block elements, find the name and chemical formula of at least one mineral containing that element.
4. *Columbium* is occasionally used as an alternative name for one of the *d*-block elements. Which?

2 Complexes

Much of the terminology used for transition metal compounds goes back to Werner's coordination theory. The development of his theory is fascinating, but there is insufficient space here to review the history in other than brief detail. At the end of the eighteenth century Tassaert prepared a compound with the empirical formula $CoCl_3(NH_3)_6$. The discovery was probably accidental but Tassaert did recognize that he had something unusual. This material was interesting because it wasn't clear why the two compounds $CoCl_3$ and NH_3, in which all the valencies seem to be satisfied, should combine together to make a material stable in its own right. Over the next century many hundreds of *d*-block metal compounds were made, but little progress was made towards achieving an understanding of the bonding in such compounds. In particular, several compounds were made containing cobalt, ammonia, and chloride (Table 2.1). At the time, compounds tended to be named after their colour or after their discoverer, a practice which would now be regarded as unsatisfactory since no indication is given as to the nature of the structure. However at the time nothing was known about the structure, so this was a perfectly reasonable policy at that time.

One notable early observation was that two compounds with *different* colours possess the *same* empirical formula: in the Werner style $CoCl_3.4NH_3$. Any description of bonding must account for this. Also, while each of the compounds in Table 2.1 contains three chloride groups, they differ in nature. This is shown by the reactions of each of these species with aqueous silver cation, Ag^+. Reaction of the first, $CoCl_3.6NH_3$, with silver cation results in *all* of the chloride anion reacting to form AgCl precipitate. However, only *two thirds* of the chloride in the purple compound $CoCl_3.5NH_3$ reacts with Ag^+ to form AgCl. Just *one third* of the chloride in the violet and green compounds $CoCl_3.4NH_3$ reacts with Ag^+ to form AgCl.

$$CoCl_3.6NH_3 + Ag^+ \rightarrow 3AgCl\downarrow$$

$$CoCl_3.5NH_3 + Ag^+ \rightarrow 2AgCl\downarrow$$

$$CoCl_3.4NH_3 + Ag^+ \rightarrow 1AgCl\downarrow$$

Table 2.1. Cobalt compounds containing chloride and ammonia.

Werner's formulation	current formulation	colour	original name
$CoCl_3.6NH_3$	$[Co(NH_3)_6]Cl_3$	yellow	luteocobaltic chloride
$CoCl_3.5NH_3$	$[Co(NH_3)_5Cl]Cl_2$	purple	purpureocobaltic chloride
$CoCl_3.4NH_3$	$[Co(NH_3)_4Cl_2]Cl$	violet	violeocobaltic chloride
$CoCl_3.4NH_3$	$[Co(NH_3)_4Cl_2]Cl$	green	praseocobaltic chloride

Observations such as these resulted in a vigorous debate involving, in particular, Werner and Jørgensen. Werner suggested that the chloride which is not precipitated out by Ag^+ is *covalently* bound to the cobalt, and those that are precipitated are *not* so bound. As ammonia is removed, one at a time, along the series $CoCl_3.6NH_3 \rightarrow CoCl_3.5NH_3 \rightarrow CoCl_3.4NH_3$, then a chloride ion replaces the ammonia and becomes covalently bound to cobalt.

It was recognized at that time that many elements possess a fixed valency. Thus, the valency of F is –1 and that of Na is +1. These values are called *primary* valencies. Other elements have more than one valence. Sulphur, for instance, has valencies of +4 and +6. Werner's idea of transition metal compounds is that in addition to the usual primary valency, metals display a *secondary* valency: defined as the number of groups covalently bonded to the metal. He noticed that cobalt is bonded to a *constant* total number of ammonia and chloride groups in the compounds in Table 2.1. Thus, in $CoCl_3.6NH_3$, the three chlorides are free to react with Ag^+, leaving six bonded ammonia groups, and so on for the other compounds in Table 2.1.

primary valency ≡ oxidation state
secondary valency ≡ coordination number

Werner's 'primary valency' is now referred to as the *oxidation state* while the 'secondary valency' is now called the *coordination number*. Werner and his colleagues realized that in their compounds, the metal is surrounded by a *fixed* number of atoms. The most common *coordination number* for transition metal compounds is six, but later on Werner recognized four coordination as well. Many examples of other coordination numbers are now known.

Werner also proposed that the six bonded groups were arranged in a symmetric fashion about the cobalt atom. Possible arrangements include a planar hexagonal structure, a trigonal prismatic structure, and an octahedral structure. The number of geometrical isomers for each of these possible arrangements is tabulated in Table 2.2. Many, many experiments over many years were followed by an analysis of the numbers of isomers found for compounds with formulae MX_5Y, MX_4Y_2, and MX_3Y_3. Based upon the numbers of *observed* isomers for each of these formulae, Werner concluded that the *most likely* arrangement is for the six groups to be bonded *octahedrally* about the metal. This conclusion is now known to be correct.

Table 2.2. Number of isomers for different geometries of six-coordinate compounds.

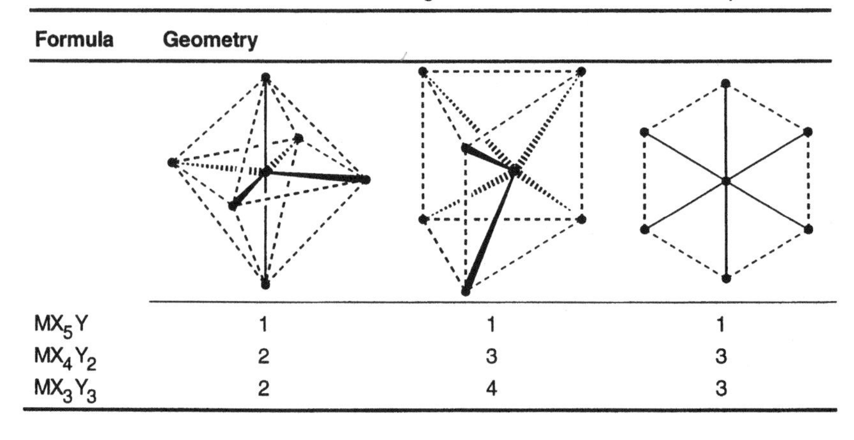

Formula	Geometry (octahedral)	Geometry (trigonal prismatic)	Geometry (planar hexagonal)
MX_5Y	1	1	1
MX_4Y_2	2	3	3
MX_3Y_3	2	4	3

2.1 A simple Lewis model of bonding

Lewis acid ≡ electron pair acceptor
Lewis base ≡ electron pair donor

Metal compounds consist of a central metal ion or atom surrounded by a set of ions or molecules which when bonded to a metal are called *ligands*. Each metal–ligand bond is a two electron interaction. Conventionally, *when a ligand is removed from a metal, it takes both the electrons in the bond with it.* If the ligand can't be removed in the test-tube, then it must be removed in a 'mental test-tube'. Under the terms of this convention, there are ligands which are removed as *neutral molecules* (H_2O, NH_3, etc.) while others are removed as *ions* (generally anions, such as Cl^- or OH^-). Since, formally, the ligands take both electrons with them, the bonding interaction holding the ligands to the metal in metal complexes is said to be *dative*, that is, the bonds are constructed by the donation of two electrons from the ligand into a bonding orbital which holds the ligand on to the metal (Fig. 2.1).

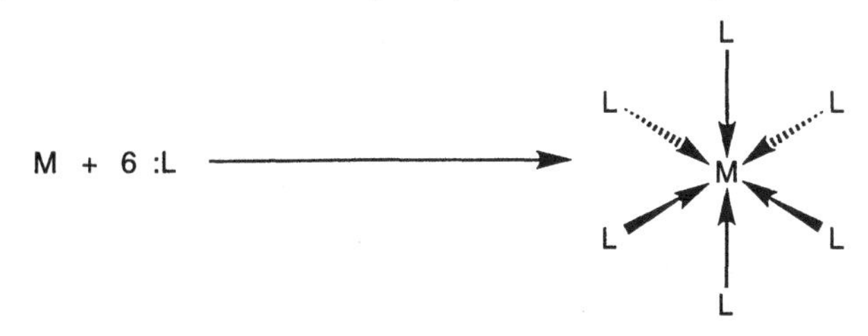

Fig. 2.1. Ligand to metal dative bonding in an octahedral system.

One good source of a pair of electrons for bonding to a metal is a lone pair. Ammonia, NH_3, or water, H_2O are ideal. The former has a single lone pair and the latter two, but normally only one of these two is accessible for bonding to a metal. Carbon monoxide, :C≡O:, has lone pairs at both ends of the molecule, but in most cases the carbon lone pair is that which binds to a metal. Ligands with one lone pair found suitable for bonding to a transition metal are termed *two-electron ligands.*

The main group chemist is familiar with *dative*, or *coordinate*, bonding. Boron trifluoride, BF_3, is a six-electron boron compound, with the boron two electrons short of the octet configuration. It is therefore a *Lewis acid.* It attains the octet configuration by acquiring *two* electrons from another molecule. Ammonia is a good electron pair donor and the nitrogen shares its lone pair with boron in the *adduct* $F_3B{\leftarrow}{:}NH_3$ on reaction with ammonia, (Fig. 2.2).

The arrow '→' is often, but not always, used to denote that one atom donates *both* electrons to that bond. A bond in which both electrons originate from one atom is a *dative covalent bond*. In this example, NH_3 is a Lewis base and BF_3 is a Lewis acid.

The origin of the term 'dative' is in the Latin *dare*, to give

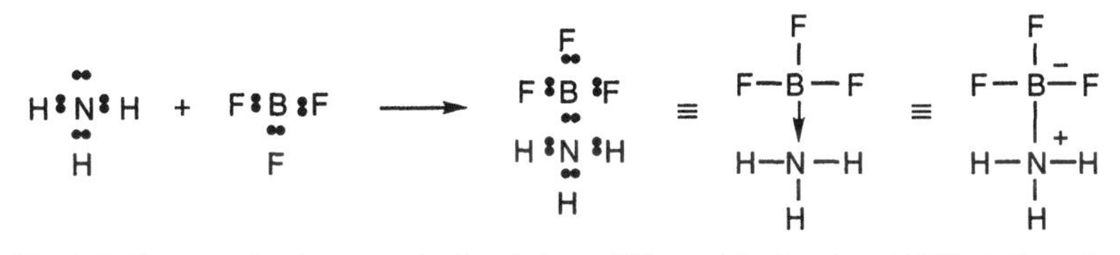

Fig. 2.2. The reaction between the Lewis base NH_3 and the Lewis acid BF_3 to form the adduct $H_3N{\rightarrow}BF_3$.

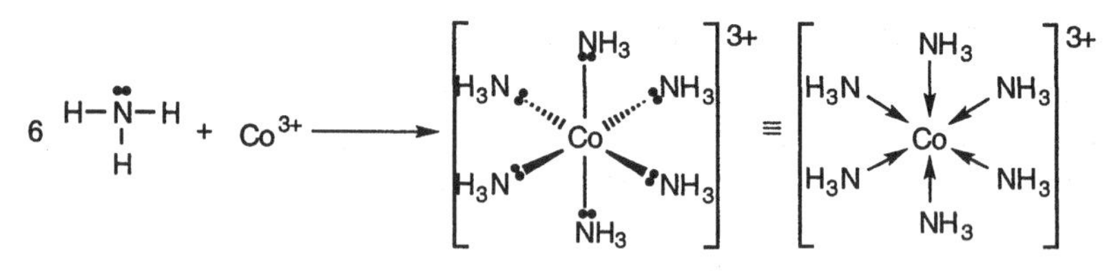

Fig 2.3. The Lewis model of $[Co(NH_3)_6]^{3+}$.

Generally the concept of complexes applies to *d*-block elements, but species such as $[AlF_6]^{3-}$ are also usually referred to as complexes

In metal complexes the metal ion functions as a Lewis acid towards a set of six ions or molecules, which are Lewis bases. These Lewis bases, by definition, possess one or more lone pairs, and when bonded to a metal ion are called *ligands*. The bonding of each ligand is *dative coordinate*, donating one or more lone pairs to the Lewis acid metal. When a metal atom or ion is bound to a set of ligands in this way, the whole is referred to as a *complex*. Using this representation, the compound $CoCl_3.6NH_3$ is now regarded as a complex involving a Co^{3+} ion and six donor ammonia ligands (Fig. 2.3). Overall, this complex possesses a +3 charge. The rôle of the three chlorides, those that react with Ag^+, is to counterbalance the +3 charge. In this scheme, Werner's primary valency (oxidation state) of cobalt is +3. The secondary valency (coordination number) is 6, the number of coordinated ammonia molecules.

Water is also a very effective ligand for many transition metal ions. The result of dissolving the metals in sulphuric acid discussed in Chapter 1 is the formation of ionic complexes with the general formula $[M(OH_2)_6]^{2+}$. The ferrous ion, Fe(II), in aqueous solution is present as a Fe(II) ion bonded though coordinate (dative) bonds to six water ligands. The structure of the complex formed when the ferrous ion is bonded to six H_2O (aqua) ligands can be represented as in the left-hand structure of Fig 2.4 in which the six arrows denote the coordinate bond. The coordination chemist normally omits the arrow heads and just draws connecting lines to represent the bonds, as in the right-hand side of Fig. 2.4.

This behaviour is displayed by any two electron donor ligands such as NH_3, Cl^-, or CO. A pictorial representation of the *localized* σ-bonding for M—NH_3, M—Cl, and M—CO complexes is illustrated in Fig. 2.5. Clearly there are parallels in the three cases.

This very simple picture has some problems. This bond model does not explain the magnetic data associated with transition metal complexes, and it does not explain their electronic spectra (that is, their colour).

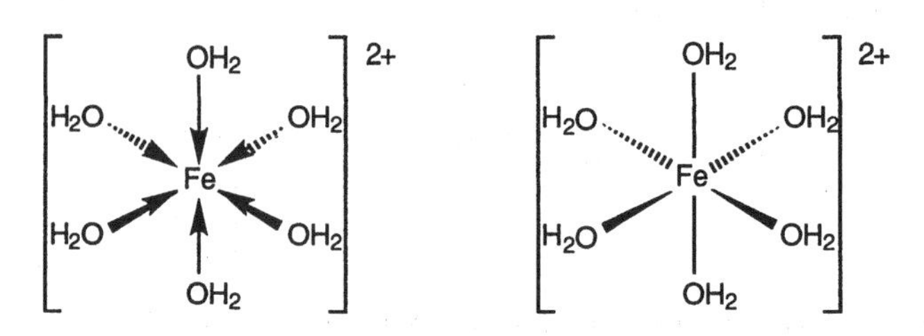

Fig. 2.4. Two ways to represent the structure of the ion $[Fe(OH_2)_6]^{2+}$.

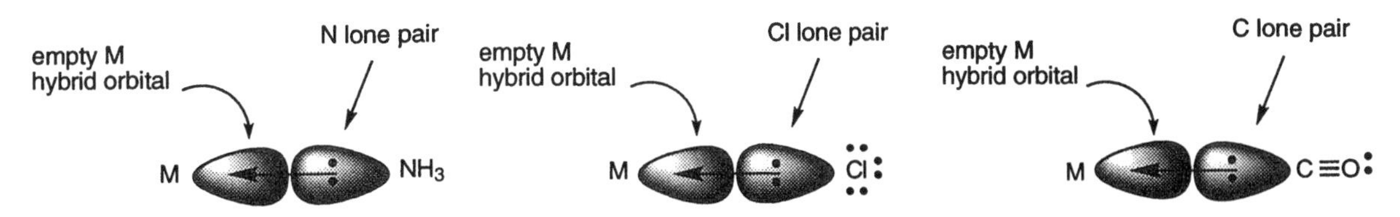

Fig. 2.5. Lone pair donation from NH_3, Cl^-, and CO into an empty metal hybrid orbital to form a dative σ-bond.

2.2 Ligands

Ligands are attached to the metal, generally via one atom (the donor atom). The donor atoms are usually elements from Groups 14 – 17. Such ligands may be simple monatomic ions such as F^- or Cl^-, or polyatomic anions such as CN^-, CH_3^- or SCN^-. Alternatively they might be neutral molecules such as NH_3 or OH_2. There are a few ligands conveniently regarded as cations, notably NO^+. Examples of typical donor atoms are tabulated in Table 2.3. Note that there is no requirement for all the ligands in a complex to be the same. It is perfectly in order for a metal complex to have a mixed ligand set.

Generally the ligand lone pair is a 'conventional' lone pair of the type found in NH_3. The two electrons located in the π-bond of ethene also function like a lone pair, but here the 'lone pair' lies between two carbon atoms, and is distributed in two sites across the C—C axis. Even this rather

The 'η^2' terminology means that the ligand uses two atoms to bond to the metal. Most ligands are bonded through a single atom and are η^1, but in such cases the η^1 label often is omitted.

Table 2.3. Examples of neutral two electron ligands arranged by position in the Periodic Table. The attached atom is written first. R typically = alkyl or aryl; X = halide.

Group 14	Group 15	Group 16	Group 17
CO, CS, C≡NR, η^2-C_2H_4, η^2-HC≡CH	NH_3, NR_3, N_2, N≡CR, Py	OH_2, OHR, OR_2, O_2, O=CMe_2, O=SMe_2	
	PX_3, PR_3, $P(OR)_3$	SH_2, SR_2	
	AsR_3	SeR_2	
	SbR_3	TeR_2	

Table 2.4. Examples of common anionic two-electron ligands. The attached atom is written first.

Group 14	Group 15	Group 16	Group 17
			H^-
CN^-, CNO^-, CNS^-, Me^-, Ph^-	NH_2^-, NO^-, NO_2^-, NCS^-	OH^-, OR^-, $OCOR^-$ ONO^-, $OClO_3^-$	F^-, FBF_3^-
SiR_3^-	PR_2^-	SH^-, SR^-, SCN^-	Cl^-
GeR_3^-			Br^-
SnR_3^-			I^-
PbR_3^-			

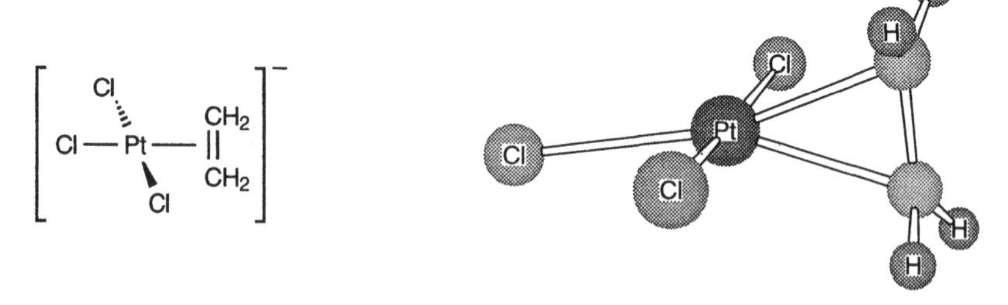

Fig 2.6. Solid state structure of $[PtCl_3(\eta^2\text{-}C_2H_4)]^-$ in $K[PtCl_3(\eta^2\text{-}C_2H_4)].(H_2O)$.

odd shaped 'lone pair' can donate its electrons to a metal. When ethene binds to a metal, the result is a triangular C_2M unit. Many alkenes are capable of binding to metals, and it is this ability that is responsible for the catalytic activation of alkenes by certain *d*-block metal compounds.

The platinum in $K[PtCl_3(\eta^2\text{-}C_2H_4)]$ *could* be regarded as five-coordinate in the sense that it is attached directly to five atoms. However, in this compound, the ethene ligand is regarded as occupying a *single* coordination site since there is a *single* ligand donor orbital. So the platinum in $K[PtCl_3(\eta^2\text{-}C_2H_4)]$ is said to be four-coordinate despite there being five directly attached atoms.

If a compound contains an M—Cl group, then by convention this is recognized as a consisting of an M^+ associated with a Cl^- ligand. Since the Cl^- is removed as a negatively charged species, it is referred to as an anionic ligand. The Cl^- ion possesses *four* lone pairs. More often than not only one of these is used in dative bonding. There are many other anionic ligands, a few examples of which are shown in Table 2.4.

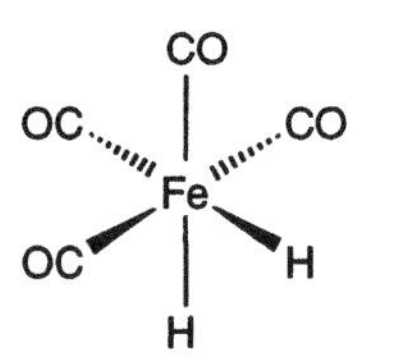

Fig. 2.7. The metal hydride complex $[FeH_2(CO)_4]$.

Perhaps the simplest anionic ligand is H^-, hydride. A metal hydride is a metal complex containing one or more hydrogen atoms directly bound to a transition metal atom. The oxidation state formalism requires that the M—H interaction be treated as hydridic, even though in many cases this is just the opposite of the chemical behaviour actually displayed (Table 2.5). Hydrides are extremely important because in many catalytic reactions it is a metal hydride that is responsible for the transfer of hydrogen atoms during catalytic cycles. The bonding is conveniently viewed as simple lone pair donation from H^- to the metal. The nature of metal hydrides remained controversial for many years. The first hydride complex discovered was $[FeH_2(CO)_4]$ (Fig. 2.7) in 1931 but it was decades later after X-ray and neutron diffraction studies that direct M—H bonds became recognized as fact.

Table 2.5. pK_a Values for some metal hydride complexes (in water).

complex	pK_a	complex	pK_a
$[HV(CO)_6]$	strong acid	$[HCo(CO)_4]$	strong acid
$[HV(CO)_5(PPh_3)]$	6.8	$[HCo(CO)_3(PPh_3)]$	6.96
$[HMn(CO)_5]$	7.1	$[HCo(CO)_3\{P(OPh)_3\}]$	4.95
$[H_2Fe(CO)_4]$	4.0	$[HFe(CO)_4]^-$	14

2.3 Inner/outer sphere complexes

The primary, or first, coordination sphere of a complex is defined as the set of ligands directly bonded to the metal atom or ion. It is possible for some groups, typically anions such as sulphate, SO_4^{2-}, to bind to a metal complex, usually positively charged, without displacing any of the ligands already in place. One such example is the interaction between $[Mn(OH_2)_6]^{2+}$ and SO_4^{2-}. The sulphate is associated with the complex cation but is *not directly bonded* to the metal. The sulphate is said to be in the *outer* coordination sphere and $[Mn(OH_2)_6]^{2+}SO_4^{2-}$ is referred to as an *outer sphere complex.* If the sulphate were to displace one water ligand, the resulting $[Mn(OH_2)_5(SO_4)]$ is an *inner sphere complex* of SO_4^{2-} with Mn(II) (Fig. 2.8).

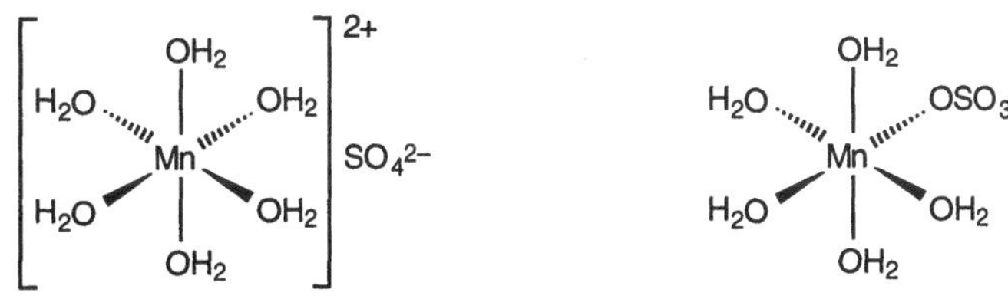

Fig 2.8. Inner and outer sphere complexes of the sulphate ligand.

2.4 Polydentate ligands

A molecule such as $H_2NCH_2CH_2NH_2$ (1,2-diaminoethane, or ethylenediamine, abbreviated en), has two nitrogen atoms with lone pairs. Both can bind to a metal. The ligand en is therefore a four-electron ligand when bonding through both nitrogen atoms. Ligands which can bind to a metal through just one lone pair are termed *monodentate* or *unidentate*. The class of ligands which can bind to a metal through two or more atoms are termed *polydentate ligands*. Ligands which bind through two atoms are *didentate*, those with three donor atoms are *tridentate*, and so on (Table 2.6).

Table 2.6. Terminology for polydentate ligands.

donor atoms	name
2	didentate
3	tridentate
4	tetradentate
5	pentadentate
6	hexadentate

Didentate ligands

In principle, the two donor atoms in $H_2NCH_2CH_2NH_2$ can both bind to the same metal, or each can bind to different metal atoms. Normally, both the donor atoms of $H_2NCH_2CH_2NH_2$ bind to the *same* metal. This results in a ring system containing the metal. When a ligand binds through two or more donor atoms to a single metal, the ligand is referred to as a *chelate*. This term is derived from the Greek *chele* (claw). When the ligand en binds to a metal, a five-membered chelate ring is formed. The saturated C and N atoms adopt a conformation which retains near tetrahedral angles in the ligand and which also achieves a N—M—N bond angle of about 90°. One effect of this is that the ring formed by the chelate complex is puckered (Fig. 2.9).

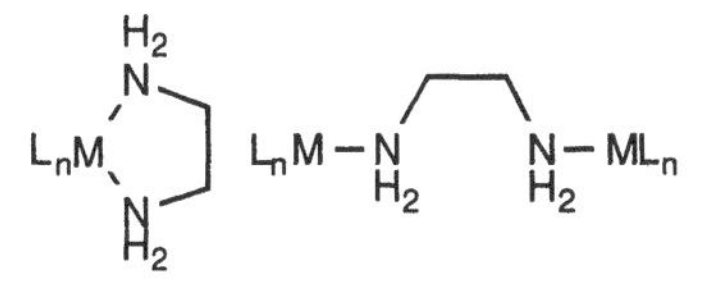

M denotes some *d*-block metal and L_n denotes the remaining ligand set.

Occasionally, some didentate ligands bridge two metals to give a dinuclear complex. In the complexes $[ClAgNH_2CH_2CH_2NH_2AgCl]$ or $[Ag_2(en)_2]^{2+}$ each nitrogen of the en ligand is coordinated to a different metal. The metal ions are *bridged* by the en ligands.

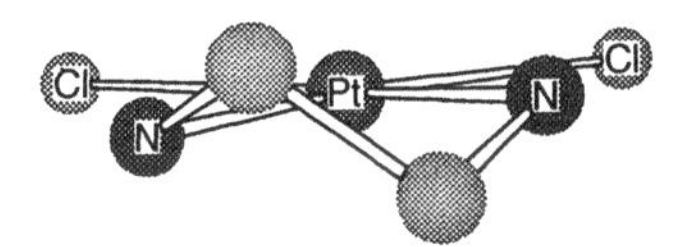

Fig. 2.9. The puckered ring in $[PtCl_2(en)]$. Hydrogen atoms omitted for clarity.

Examples of other neutral polydentate ligands (Fig. 2.10) include compounds such as $Ph_2PCH_2CH_2PPh_2$ (a four-electron ligand when both P atoms are bonded to the metal, often abbreviated as diphos or dppe) and

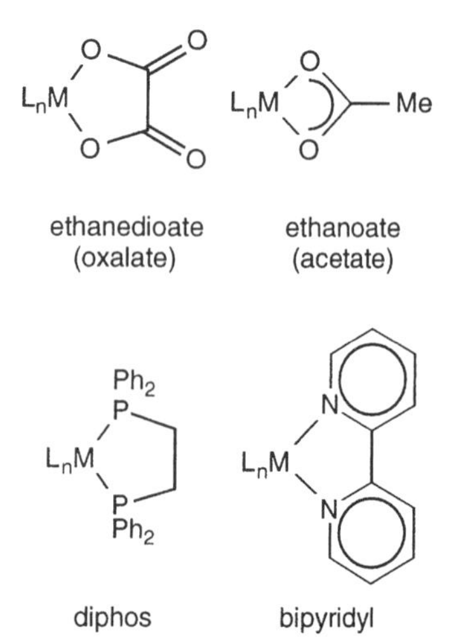

Fig. 2.10. Examples of some didentate ligand interactions with a metal.

bipyridyl, (bpy, a four-electron ligand bonded through both N atoms). Examples of anionic polydentate ligands include ethanoate (acetate, $MeCO_2$), a didentate four electron donor when bonded through both oxygen atoms) and oxalate ($C_2O_4^{2-}$, a didentate ligand bonded through two of the four oxygen atoms). A number of other common anions sometimes act as chelates, forming four-membered rings with the metal. Examples include nitrate (NO_3^-), sulphate (SO_4^{2-}) and perchlorate (ClO_4^-).

One notable feature of chelating ligands is their ability to bind very strongly to metal cations. When the resulting complex is overall electrically neutral, the complex often has good solubility in organic solvents, whereas complexes with an overall charge tend to be water soluble. This is the basis for the extraction of metal ions from aqueous media into organic solvents. The ligand H_2Dz (dithizone) is a good example. Under certain conditions it forms strong complexes with certain metal cations and the resulting *neutral* complexes, such as $[Ni(HDz)_2]$ (Fig. 2.11) are extractable into organic solvents such as dichloromethane They also often form highly coloured complexes and form the basis of a colorimetric assay of metal ions.

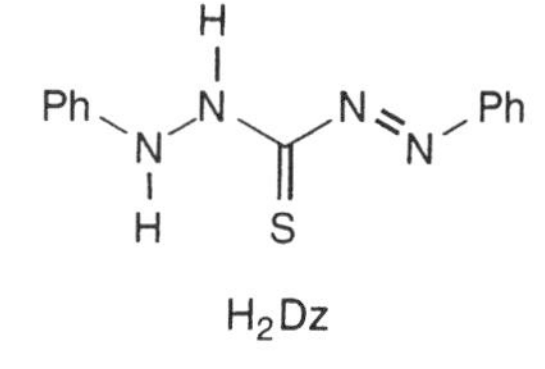

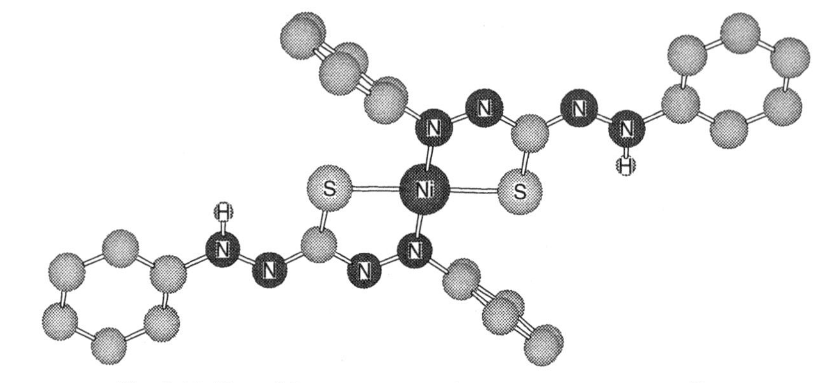

Fig. 2.11. The solid state structure of the Ni(II) complex of HDz^-.

A chelating ligand known as an oxime (Fig. 2.12) forms a neutral complex with Pd(II). Since the resulting complex is neutral, it can then be extracted (and so concentrated) into an organic medium. This reaction is one component in an industrial process for the separation of the platinum group metals by means of solvent extraction.

An alkene is able to donate two electrons from the π bond to the metal. A *diene* such as butadiene donates a total of four electrons from the two π bonds, and is therefore a didentate ligand. A good example is

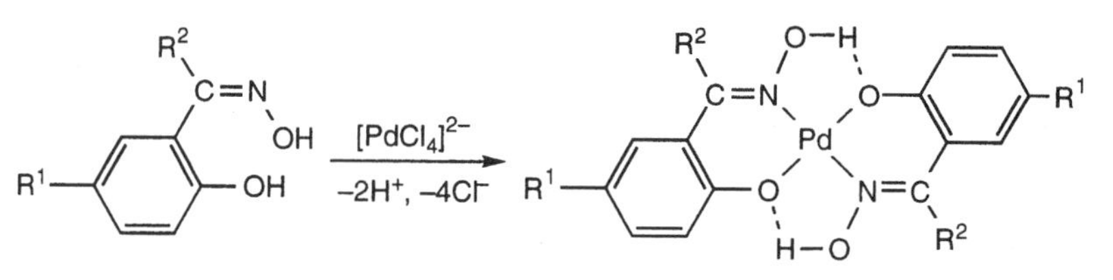

Fig. 2.12. The formation of an oxime complex which is soluble in organic solvents from $[PdCl_4]^{2-}$. R^1 = alkyl; R^2 = alkyl or aryl.

$[Fe(CO)_3(\eta^4\text{-}C_4H_6)]$ (Fig. 2.13, the η^4 nomenclature indicates that the number of bonded atoms in the butadiene ligand is 4) in which the butadiene ligand occupies two sites in the five coordinate iron complex. In one sense this complex might be regarded as seven-coordinate. However, recall that ethene is regarded as occupying a single coordination site (Section 2.2). By analogy a diene occupies two coordination sites and the iron in $[Fe(CO)_3(\eta^4\text{-}C_4H_6)]$ is said to be five-coordinate.

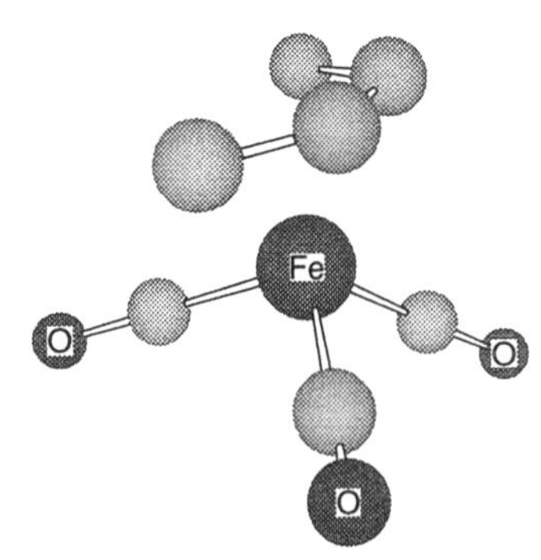

Fig. 2.13. The structure of $[Fe(CO)_3(\eta^4\text{-}C_4H_6)]$.

Tridentate ligands

Tridentate ligands use *three* donor atoms in bonding to the metal ion. Examples are given in Fig 2.14.

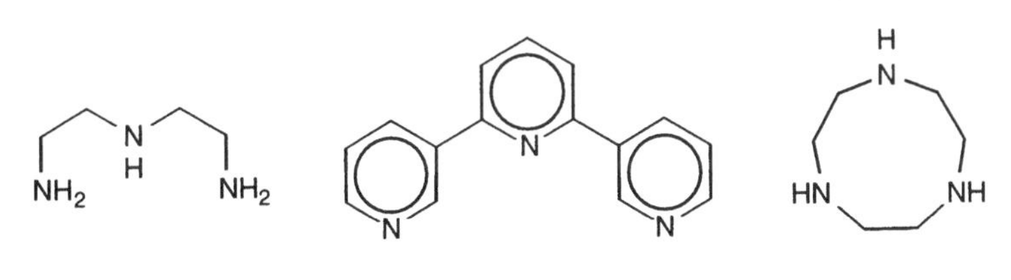

Fig. 2.14. Examples of tridentate ligands.

The last of these, 1,4,7-triazacyclononane is an example of a special class of ligands with a *heterocyclic* framework known as *macrocyclic* ligands. Macrocyclic ligands consist of a ring of at least nine atoms in which there are at least three donor atoms. Such ligands often form particularly stable complexes with transition metals. They are important in biological and industrial chemistry.

Certain polyalkenes are also capable of binding to three coordination sites on a metal. Cycloheptatriene has six π-electrons in three double bonds. Each double bond is capable of binding to metal, as in $[Cr(CO)_3(\eta^6\text{-}C_7H_8)]$. The chromium in this complex is regarded as six coordinate rather than nine-coordinate since each double bond occupies just one vertex (Section 2.2).

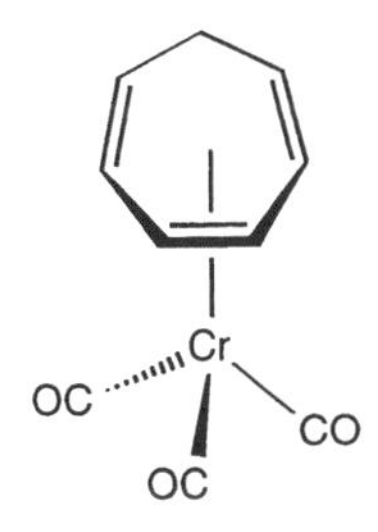

Arenes such as benzene, C_6H_6, also possess six π-electrons. Both are capable of occupying *three* coordination sites of a metal and are particularly important in the chemistry of low oxidation state metal compounds. The classic examples of benzene complexes are $[Cr(CO)_3(\eta^6\text{-}C_6H_6)]$ and $[Cr(\eta^6\text{-}C_6H_6)_2]$ (Fig. 2.15). The arene is regarded as occupying three coordination sites and is a six electron donor. Each of the carbon aromatic π-electrons is therefore involved in this bonding.

Another organic species with six π-electrons is $[C_5H_5]^-$, the

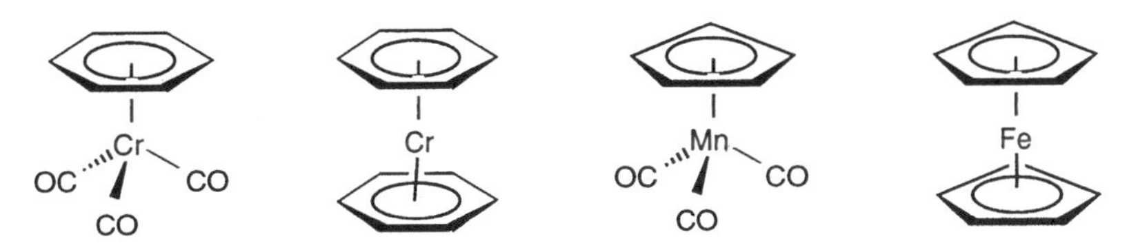

Fig. 2.15. Examples of benzene and cyclopentadienyl complexes.

cyclopentadienyl ligand. The manganese complex $[Mn(CO)_3(\eta^5\text{-}C_5H_5)]$ (Fig. 2.14) is a good example. Again the ligand is regarded as occupying three coordination sites. Now the manganese is formally in oxidation state +1, whereas in both chromium complexes the metal is in the zero oxidation state. This is because for electron counting purposes the five membered ring must come off in its closed shell configuration, $[C_5H_5]^-$ (which has *six π*-electrons). Perhaps the most famous cyclopentadienyl complex is $[Fe(\eta^5\text{-}C_5H_5)_2]$, otherwise known as *ferrocene* (Fig. 2.14). Ferrocene and dibenzene chromium are examples of 'sandwich' complexes, the metal ion being 'sandwiched' between two planar cyclic molecules. The iron in ferrocene is present as Fe(II). The cyclopentadienyl ligand is so common that it is given equal status with groups such as Me and Ph in that it has its own special abbreviation, Cp. Thus, ferrocene is often written as $[FeCp_2]$.

Tetradentate ligands

Tetradentate ligands use four donor atoms in bonding to the metal ion. Some examples are shown in Fig. 2.16.

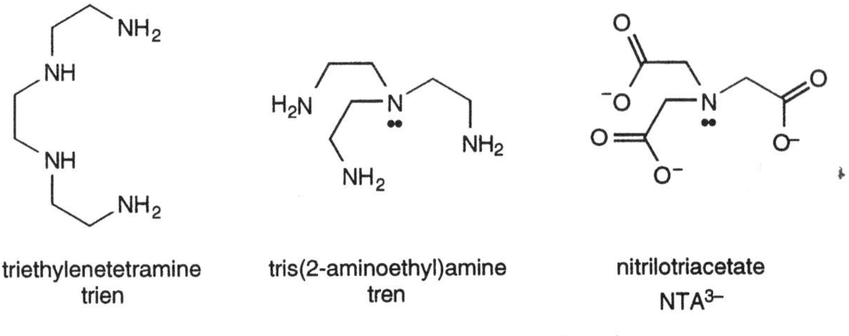

Fig. 2.16. Examples of tetradentate ligands.

There are many important tetradentate macrocyclic ligands. Iron complexes of the *porphyrin* ring (Fig. 2.17) are present in proteins such as haemoglobin and myoglobin which are involved in the transfer and storage of dioxygen. The *corrin* ring is the cobalt binding function in vitamin B_{12} (Fig. 1.4).

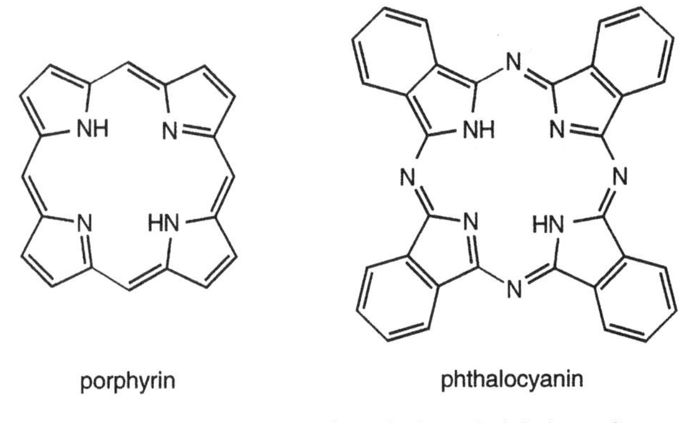

Fig. 2.17. The structures of porphyrin and phthalocyanin.

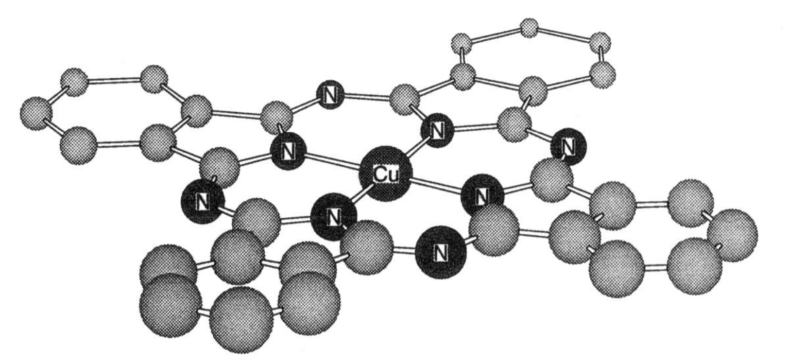

Fig. 2.18. The solid state structure of a copper phthalocyanin complex. Note the planar CuN_4 unit.

Phthalocyanines are synthetic macrocycles which form highly coloured and commercially important metal complexes isolated as intensely coloured metal salts. These are planar (Fig. 2.17). Note that in these cases placing a metal cation (typically M^{2+}) into the porphyrin and phthalocyanin rings requires that the two internal hydrogen atoms are removed as protons in order for complexation to proceed (Fig. 2.18).

Pentadentate and hexadentate ligands

Pentadentate ligands are rare but one particularly important *hexadentate* ligand is the tetraanion of H_4-edta (ethylenediaminetetraacetic acid, Fig. 2.19). This anion is important in analytical chemistry and medicine because of its ability to bind strongly to metals. Sometimes edta is *pentadentate*, in which case it is a trianion (H-edta^{3-}), as one of the acid groups is not ionized (Fig. 2.20). The sixth site is then occupied by H_2O. In addition to having the ability to bind to the *d*-block metals, the edta tetraanion forms very strong complexes with main group metal cations in which the ligand very effectively wraps the metal up, that is, the ligand 'sequesters' the metal ion. This strong complexation is the basis of the use of edta as a water softener. Water is softened by edta since edta forms water soluble complexes with Ca^{2+} and Mg^{2+} ions, preventing their precipitation by soaps.

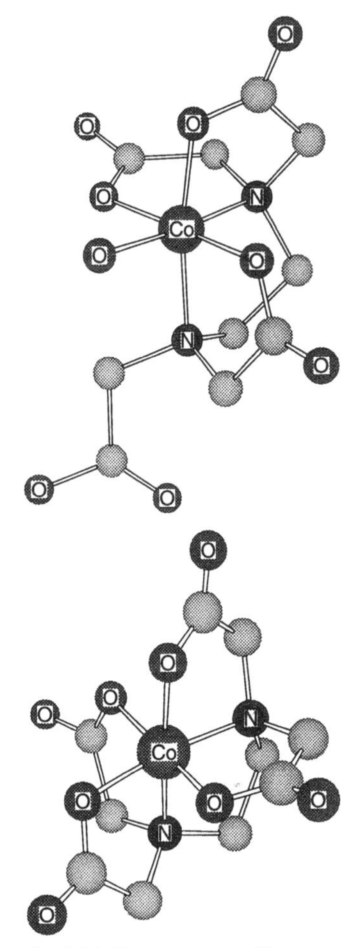

Fig. 2.20. The structures (H atoms excluded) of the neutral Co(III) complex of H-edta^{3-} ([Co(OH_2)(H-edta)], top) and the monoanionic Co(III) complex of edta^{4-} (NH_4[Co(edta)], bottom).

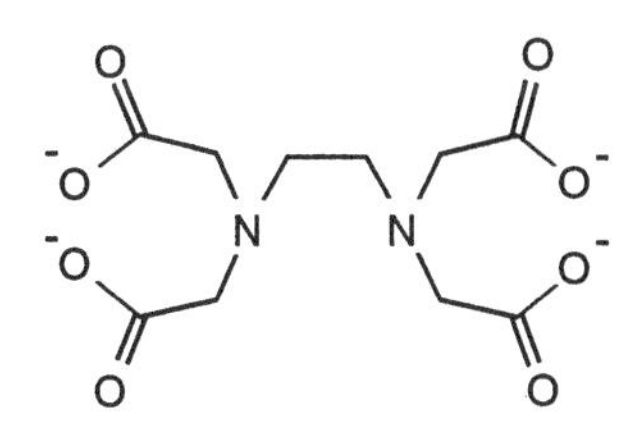

Fig. 2.19. The structure of the edta tetraanion.

2.5 Stability constants

Some ligands form stronger bonds with metals than others. A common reaction of transition metal compounds is *ligand substitution* in which one or more of the current set of ligands is replaced by new ligands. It is useful to have numerical data with which to discuss such phenomena.

In aqueous solution, metal ions, such as the Fe(II) ion, exist as metal complexes in which water molecules are bound to the Fe(II) ion: $[Fe(OH_2)_6]^{2+}$ in this case. Ligand substitution by cyanide ion for a water ligand in $[Fe(OH_2)_6]^{2+}$ gives a cyano complex (Eqn 2.1).

$$[Fe(OH_2)_6]^{2+} + CN^- \rightleftharpoons [Fe(OH_2)_5(CN)]^+ + H_2O \qquad (2.1)$$

This is an equilibrium process which has an equilibrium constant associated with it. The value of the constant is given by Eqn 2.2 in which each term refers to the concentration of that species. It is a little unfortunate that square brackets are taken to represent a metal complex *and* to represent the concentration of whatever species is contained within the square brackets. Here the square brackets denoting a complex are omitted and the square brackets should be taken to denote concentration.

$$\text{constant} = \frac{[Fe(OH_2)_5(CN)^+][H_2O]}{[Fe(OH_2)_6{}^{2+}][CN^-]} \qquad (2.2)$$

For this reaction, the value of the constant is much greater than 1, so the equilibrium lies over to the right, that is, the reaction tends to proceed towards the species on the right. In practice, this reaction might be performed in aqueous solution. The water eliminated will make almost no difference to the bulk water concentration $[H_2O]$, meaning that the water concentration remains constant throughout. Dividing both sides of the equation through by the concentration of water, $[H_2O]$, gives a new equation for which the value of constant/$[H_2O]$ is a new constant, labelled K_1 (Eqn 2.3).

$$K_1 = \frac{\text{constant}}{[H_2O]} = \frac{[Fe(OH_2)_5(CN)^+]}{[Fe(OH_2)_6{}^{2+}][CN^-]} \qquad (2.3)$$

The reaction does not stop after this substitution and a series of further substitutions proceeds until all the water ligands are replaced by cyanide. Replacement of the second water ligand gives a neutral complex (Eqn 2.4). This is again an equilibrium process and an analogous equation to Eqn 2.2 may be written to express this (Eqn. 2.5).

$$[Fe(OH_2)_5(CN)]^+ + CN^- \rightleftharpoons [Fe(OH_2)_4(CN)_2] + H_2O \qquad (2.4)$$

$$K_2 = \frac{[Fe(OH_2)_4(CN)_2]}{[Fe(OH_2)_5(CN)^+][CN^-]} \qquad (2.5)$$

Now, since both K_1 and K_2 are constant, the product K_1K_2 is also a constant (Eqn. 2.6). This new constant is called β_2. After cancellation and rearrangement leads to Eqn. 2.7.

$$\beta_2 = K_1K_2 = \frac{[Fe(OH_2)_5(CN)^+]}{[Fe(OH_2)_6{}^{2+}][CN^-]} \times \frac{[Fe(OH_2)_4(CN)_2]}{[Fe(OH_2)_5(CN)^+][CN^-]} \qquad (2.6)$$

$$\beta_2 = \frac{[Fe(OH_2)_4(CN)_2]}{[Fe(OH_2)_6{}^{2+}][CN^-]^2} \qquad (2.7)$$

The constant β_2 is called an overall *stability constant* and refers to the chemical equation (Eqn 2.8) expressing the overall replacement of two water ligands.

$$[Fe(OH_2)_6]^+ + 2CN^- \rightleftharpoons [Fe(OH_2)_4(CN)_2] + 2H_2O \qquad (2.8)$$

Similar equations may be written down for each of the remaining substitution processes up to β_6.

$$[Fe(OH_2)_4(CN)_2] + CN^- \rightleftharpoons [Fe(OH_2)_3(CN)_3]^- + H_2O$$

$$\beta_3 = \frac{[Fe(OH_2)_3(CN)_3{}^-]}{[Fe(OH_2)_6{}^{2+}][CN^-]^3}$$

$$[Fe(OH_2)_3(CN)_3]^- + CN^- \rightleftharpoons [Fe(OH_2)_3(CN)_3]^{2-} + H_2O$$

$$\beta_4 = \frac{[Fe(OH_2)_2(CN)_4{}^{2-}]}{[Fe(OH_2)_6{}^{2+}][CN^-]^4}$$

$$[Fe(OH_2)_2(CN)_4]^{2-} + CN^- \rightleftharpoons [Fe(OH_2)(CN)_5]^{3-} + H_2O$$

$$\beta_5 = \frac{[Fe(OH_2)(CN)_5{}^{3-}]}{[Fe(OH_2)_6{}^{2+}][CN^-]^5}$$

$$[Fe(OH_2)(CN)_5]^{3-} + CN^- \rightleftharpoons [Fe(CN)_6]^{4-} + H_2O$$

$$\beta_6 = \frac{[Fe(CN)_6{}^{4-}]}{[Fe(OH_2)_6{}^{2+}][CN^-]^6}$$

For any given reaction and for an arbitrary number of ligands n, the form of the equation for β_n is given by Eqn 2.9.

$$\beta_n = K_1 \times K_2 \times K_3 \times \ldots K_n = \frac{[ML_n]}{[M][L]^n} \qquad (2.9)$$

The quantity β_6 is the *stability constant* for the complex $[Fe(CN)_6]^{4-}$ and its numerical value is about 10^{35}, meaning that the series of equilibria favour very much the hexacyano ion $[Fe(CN)_6]^{4-}$. It is quite common to express the log values of any β_n rather than the actual value of β_n. This is because the value of $\log(\beta_n) = 35$ is much more manageable in discussions than the number 10^{35}. Note that for any given series of equilibrium constants K_1, K_2, K_3, ..., K_n, each value of K tends to be smaller than the preceding value, that is, $K_2 < K_1$, $K_3 < K_2$, and so on.

The stability constant is also called the overall formation constant

2.6 Chelate effect

The chelate effect is demonstrated by comparing the reaction of a chelating ligand and a metal ion with the corresponding reaction involving comparable but unidentate ligands. In this sense, the ligand en is equivalent to two NH_3 ligands. The reactions of interest are represented by Eqns 2.10 and 2.11.

$$[M(OH_2)_6]^{2+} + 6\,NH_3 \rightleftharpoons [M(NH_3)_6]^{2+} + 6\,H_2O \qquad (2.10)$$

$$[M(OH_2)_6]^{2+} + 3\,en \rightleftharpoons [M(en)_3]^{2+} + 6\,H_2O \qquad (2.11)$$

In practice, the equilibrium constants for reaction 2.11 are larger than those for reaction 2.10. The equilibrium constant of a reaction is related to the standard free energy of a reaction by Eqn 2.12. Inspection of this equation shows that reactions for which K is large (that is, the reaction proceeds to the right) have a large *negative* value of ΔG. This means that the value of $\Delta G°$ is much more negative for reaction 2.11 than for reaction 2.10. The free energy change is also related (Eqn 2.12) to the enthalpy and entropy changes ($\Delta H°$ and $\Delta S°$ respectively) and the temperature, T (degrees Kelvin).

$$\Delta G° = -RT\ln K = \Delta H° - T\Delta S° \qquad (2.12)$$

The ligand en makes two M—N bonds in the M(en) unit which are comparable with the two M—N bonds in a $M(NH_3)_2$ unit. In practice, the similarity of these M—N interactions means that the *enthalpy* changes in reactions such as those in Eqns 2.10 and 2.11 are rather similar. Therefore the differences in observed values of $\Delta G°$ must be related to the *entropy* change. Notice that the reaction in Eqn 2.10 involves seven species (the initial complex and the six ammonia molecules) being converted into seven new species (the final product and the six eliminated water molecules). This, therefore, should not result in a particularly large entropy change. However the reaction in Eqn 2.11 is different. Here four species (the initial complex and the three en molecules) are converted into seven new species (the final product and the six eliminated water molecules). This represents an increase in disorder in the system, that is, an increase in entropy. This means that the $-T\Delta S°$ term makes a large negative contribution to $\Delta G°$, and results in a larger equilibrium constant.

All this is conveniently illustrated with values for the first stages of equations 2.10 and 2.11 involving Cu(II) (Eqns 2.13 and 2.14).

$$[Cu(OH_2)_6]^{2+} + 2\,NH_3 \rightleftharpoons [Cu(OH_2)_4(NH_3)_2]^{+} + 2\,H_2O \qquad (2.13)$$

$\beta_2 = 10^{7.7}$; $\Delta H° = -46$ kJ mol^{-1}; $\Delta S° = -8.4$ J K^{-1} mol^{-1}

$$[Cu(OH_2)_6]^{2+} + en \rightleftharpoons [Cu(OH_2)_4(en)]^{2+} + 2\,H_2O \qquad (2.14)$$

$\beta_1 = 10^{10.6}$; $\Delta H° = -54$ kJ mol^{-1}; $\Delta S° = +23$ J K^{-1} mol^{-1}

The values of $\Delta H°$ are both negative showing that heat is evolved. The percentage difference in the two values is not large. The value for the en reaction is slightly larger since en interacts a little more strongly with the metal than ammonia. The differences in values for $\Delta S°$ are more dramatic. The signs for each reaction are different. The value for the en reaction is large and negative, indicating an increase in entropy, and hence a more favoured reaction. The chelate effect is also evident in reaction 2.15 in which six NH_3 ligands are replaced by three en ligands

$$[Ni(NH_3)_6]^{2+} + 3\ en \rightleftharpoons [Ni(en)_3]^{2+} + 6\ NH_3 \qquad (2.15)$$

$\Delta G° = -54$ kJ mol^{-1}; $\Delta H° = -29$ kJ mol^{-1}; $\Delta S° = +88$ J K^{-1} mol^{-1}

The value of $\Delta G°$ is negative indicating that the reaction proceeds. About half of the change in $\Delta G°$ is contributed by the enthalpy change (it seems the Ni—N interactions are slightly stronger in Ni(en) than in $Ni(NH_3)_2$. The remainder is accounted for by the large change in entropy (four species are converted into seven species).

The size of the resulting chelate ring in a complex is also a factor in the chelate effect. In general, for bis-amino complexes in which the two nitrogen atoms are linked by an aliphatic chain, the formation of a five membered chelate ring is more favourable than formation of a six membered ring, which in turn is more favourable than formation of a seven membered ring. Thus, complexes of en ($NH_2CH_2CH_2NH_2$) are more stable than complexes of 1,3-diaminopropane ($NH_2CH_2CH_2CH_2NH_2$).

2.7 The concept of hard and soft acids and bases

One important type of reaction displayed by metal complexes is the replacement of some or all of the ligand set by other ligands. A general case of this reaction is given in Eqn 2.16.

$$M—L + L^* \rightleftharpoons M—L^* + L \qquad (2.16)$$

For any given example of this reaction, if the reaction tends towards the right, then the ligand L* (a Lewis base) makes a more stable complex with the particular metal M (a Lewis acid) than does the ligand L. In this way it is possible to construct a ranking of ligand Lewis base strengths. It turns out that the ordering of ligands in this sequence is dependent upon the metal, M. Ligands such as NH_3, OH_2, OH^-, CO_3^{2-}, and SO_4^{2-}, tend to form strong complexes with metals such as Mn^{2+} or Cr^{3+}. Ligands such as CO, C_2H_4, or SRH tend to form rather weak complexes with these metals. On the other hand, metals such as Cu^+ and Ag^+ form strong complexes with CO, C_2H_4, or SRH but weaker complexes with ligands such as NH_3, OH_2, OH^-, CO_3^{2-}, and SO_4^{2-}.

The donor atoms in the ligands which form the most stable complexes with metals such as Mn^{2+} or Cr^{3+} tend to be small, electronegative, and only slightly polarizable. They are also difficult to oxidize. These are defined as *hard* bases. The metals with which they form the most stable complexes are called *hard* metals. Such metals are also small and not very polarizable.

The donor atoms of ligands which tend to form the most stable complexes with metals such as Cu^+ and Ag^+ are larger, less electronegative, and highly polarizable. They are easier to oxidize and are referred to as *soft* bases. The metals with which they interact best are called *soft* metals. These tend to be larger and more polarizable than hard metals. Some examples of each are

Table 2.7. Some hard and soft acids and bases.

	Hard	Borderline	Soft
acids	H^+, Sc^{3+}, Cr^{3+}, Cr^{6+}, MoO^{3+}, Ti^{4+}, Zr^{4+}, Hf^{4+}, VO^{2+}, Mn^+, Mn^{7+}, Fe^{3+}, Co^{3+}	Fe^{2+} Co^{2+} Ni^{2+}, Cu^{2+}, Zn^{2+}, Rh^{3+}, Ir^{3+}, Ru^{3+}, Os^{2+}	Cu^+, Ag^+, Au^+, Cd^{2+}, Hg^+, Hg^{2+}, Pt^{2+}, Pd^{2+}, all *d*-block metals in zero oxidation state
bases	NH_3, NH_2R, N_2H_4, OH_2, OH^-, O^{2-}, OHR, OR^-, OR_2, CO_3^{2-}, SO_4^{2-}, $OClO_3^-$, Cl^-, F^-, NO_3^-, PO_4^{3-}, $OCOMe^-$	NH_2Ph, N_3^-, N_2, NO_2^-, SO_3^{2-}, Br^-	H^-, R^-, C_2H_4, C_6H_6, CN^-, CO, SCN^-, PR_3, $P(OR)_3$, AsR_3, SR_2, SHR, SR^-, I^-

tabulated in Table 2.7. Note that this classification also applies to elements other than the *d*-block elements. Thus, Al^{3+} is hard while Tl^+ is soft. These observations are summarized as follows: hard acids prefer hard bases and soft acids prefer soft bases.

The concept is put on a numerical footing by defining as references a hard acid (H^+) and a soft acid ($MeHg^+$), and then measuring the tendency for bases to react with each of these acids. Consider Eqn 2.17. The position of the equilibrium in aqueous solution determines the relative affinities of the base B for the hard acid H^+ and the soft acid $MeHg^+$. Soft bases prefer to react with $MeHg^+$ rather than H^+ and so the equilibrium lies to the right. The converse is true for hard bases.

$$BH^+ + HgMe(OH_2)^+ \rightleftharpoons H_3O^+ + HgMe(B)^+ \qquad (2.17)$$

The terms hard and soft are very imprecise but do provide a 'rule of thumb' for guessing complex stability. As might be anticipated there is a whole spectrum of hardness and softness, and further there are plenty of cases where soft acids complex with hard bases and vice versa. Acids and bases which are somewhat intermediate in nature are referred to as *borderline*. Typical borderline bases include unsaturated nitrogen donor ligands.

A perusal of the literature shows that it is not particularly clear what factors are responsible for the hard–hard, soft–soft effect. This is probably because there are several contributor factors which are difficult to unravel. One approach is to regard hard–hard interactions as largely electrostatic, or ionic. Soft–soft interactions, on the other hand, are largely covalent. It seems that π-bonding (covalent) effects are often important in soft–soft interactions.

2.8 Exercises

1. For each of the ligands in this chapter, find the structure of two complexes of that ligand. For each, describe the geometric arrangement described by the set of atoms directly bonded to the central metal.
2. Metals are sometimes categorized as type *a* or type *b* metals rather than as *hard* or *soft*. Use your other text books to compare the definitions of types *a* and *b* metals with *hard* and *soft* metals.

3 Shape and isomerism

It is common for the novice to be confused by the apparently bewildering array of structural types available for metal complexes as compared to the limited numbers of available 'coordination' types available for organic compounds. There are well defined rules for the prediction of geometry in *p*-block chemistry such as the VSEPR model. With time the reader will recognize that there are rules for the prediction of metal structural types, it's just that there is more variation possible for metal compound structures.

A *d*-block metal has a total of nine atomic orbitals. Some of these are used to form bonds with ligands, while others are nonbonding. The nonbonding electrons are effectively lone pairs, but whereas lone pairs in main group compounds tend to be directional (hence the VSEPR Rules), those in transition metal complexes exert much less influence on geometry.

It is quite difficult to use the VSEPR electron counting rules for transition metal complexes, although some of the principles can be applied. Generally, the shape of a complex is dictated by the number of coordinated atoms. The geometry of very many coordination complexes is that in which the ligands are set as far apart as possible from each other. Thus, a transition metal complex with just two ligands such as $[AuCl(PPh_3)]$ is generally linear. Complexes with three ligands are normally trigonal. Complexes with four, five, and six ligands are generally tetrahedral (but there is also an important class of square planar complexes), trigonal bipyramidal, and octahedral respectively. There are important exceptions to these generalizations, but more often than not it is the simplest geometries that are displayed.

The most common coordination numbers for the first transition metal series are *four* (tetrahedral and square planar) and *six* (octahedral). Examples of coordination numbers are in Table 3.1. Higher coordination numbers are possible for the metals in the early second and third series as these are larger.

Table 3.1. Some examples of coordination number.

complex	coordination number
$[Cu(C_6H_2Ph_3\text{-}2,4,6)]$	1
$[Cu(NH_3)_2]^+$	2
$[Pt(PPh_3)_3]$	3
$[CoCl_4]^-$	4
$[CuCl_5]^{3-}$	5
$[Co(NH_3)_6]^{3+}$	6
$[Co(en)_3]^{3+}$	6 (en is a didentate ligand)
$[Co(en)_2Cl_2]^+$	6
$[NbOF_6]^{3-}$	7
$[Mo(CN)_8]^{4-}$	8
$[ReH_9]^{2-}$	9

It is quite common for the coordination number of a metal to change during a reaction. For instance, four-coordinate square planar complexes often add two ligands to become six coordinate octahedral.

$$MA_4 + 2B \rightarrow MA_4B_2$$

The tetrahedral complex $[Pt(PPh_3)_4]$ is a crowded molecule because of the steric bulk of the phosphine ligands. As a result, $[Pt(PPh_3)_4]$ is in equilibrium with the trigonal planar complex $[Pt(PPh_3)_3]$ and free PPh_3.

$$[Pt(PPh_3)_4] \rightleftharpoons [Pt(PPh_3)_3] + PPh_3$$

The number of coordinated atoms within the primary coordination sphere of a metal complex varies from 1 to 12 (examples with the highest and lowest coordination numbers are rare). The *coordination number* is defined as the number of ligand *atoms* (not ligands) which are bonded to a metal.

3.1 Coordination number 1

There are probably no genuine examples of coordination number 1. This is because under normal circumstances such a coordination number would leave the metal very exposed to attack by other ligands, so increasing the coordination number. However, in principle, if the single ligand has a structure which prevents the approach of another ligand through steric shielding and it is kept in the correct environment, then such a complex might be sufficiently stable to isolate. One claimed example is the complex $[Cu(C_6H_2Ph_3\text{-}2,4,6)]$, but it now appears that the claim was made in error.

3.2 Coordination number 2

There are relatively few complexes with coordination number 2. As for coordination number 1, the problem is that metals with just two ligands tend to be rather exposed to attack by other ligands, so increasing the coordination

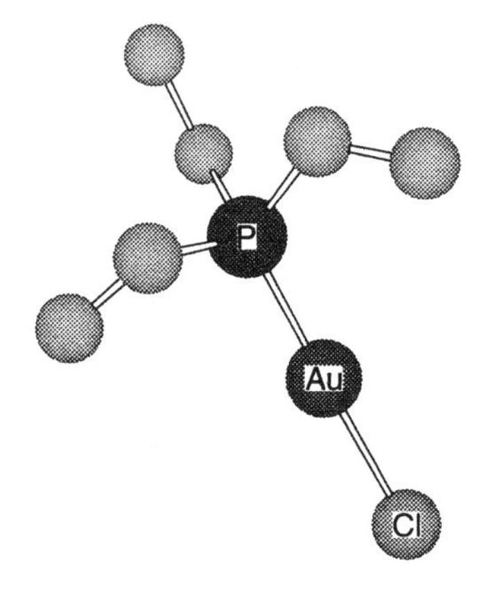

Fig. 3.1. The solid state structures of the two-coordinate complex $[AuCl(PEt_3)]$.

number. Most of those known are M(I) complexes of Group 11 metals and examples include [AuCl(PEt$_3$)] (Fig 3.1), $[CuCl_2]^-$, $[AgCl_2]^-$, $[Ag(CN)_2]^-$, and $[Au(CN)_2]^-$. Others include complexes of Hg(II) such as $[Hg(CN)_2]$ and $[HgMe_2]$. When using a hybridization view of bonding in linear systems such as this, an *sp* metal configuration is appropriate.

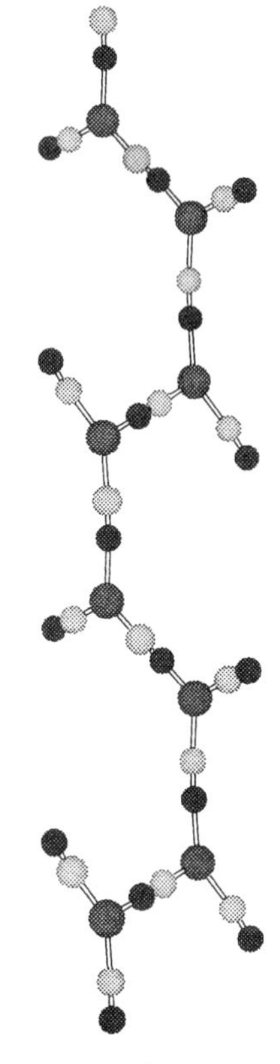

Fig. 3.4. The chain structure of $[Bu_4N][Cu(CN)_2]$.

3.3 Coordination number 3

Examples of this coordination number are also few and far between. The geometry is approximately trigonal in most of those known and corresponds to metal sp^2 hybridization. In many cases, the metal is prevented from bonding to more than three ligands by the steric shielding of the ligands (Fig. 3.2). The platinum complex $[Pt(PPh_3)_3]$ (page 23) is one such example. There are a number of complexes with the general formula $[M\{N(SiMe_3)_2\}_3]$ and the iron compound $[Fe\{N(SiMe_3)_2\}_3]$ is a crystallographically characterized example (Fig. 3.2) of a three-coordinate complex.

The salt $NBu_4[Cu(CN)_2]$ is an interesting compound. In the solid state (Fig. 3.4) the coordination geometry is not linear as might have been anticipated from the empirical formula. Instead, the anion has a chain-like structure in which the copper atoms are trigonally coordinated, each by two carbon atoms and one nitrogen atom from a total of three cyanide ligands. This is an example of a complex in which ligands *bridge* two metals. As such this represents a good example showing the danger of extrapolating the solid state coordination number from the stoichiometry.

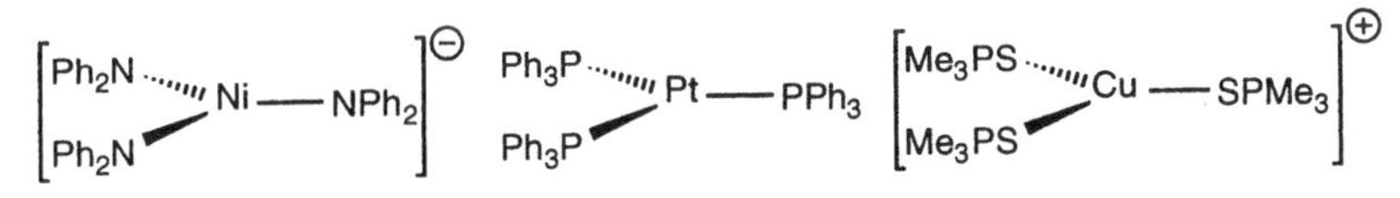

Fig 3.2. Examples of three-coordinate complexes.

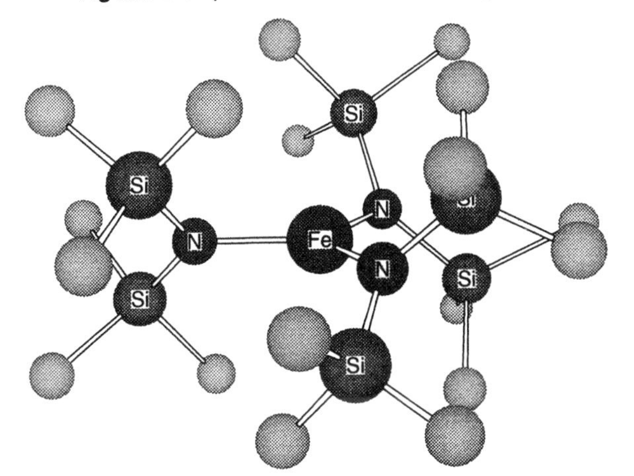

Fig 3.3. The solid state structures of the three-coordinate complex $[Fe\{N(SiMe_3)_2\}_3]$.

3.4 Coordination number 4

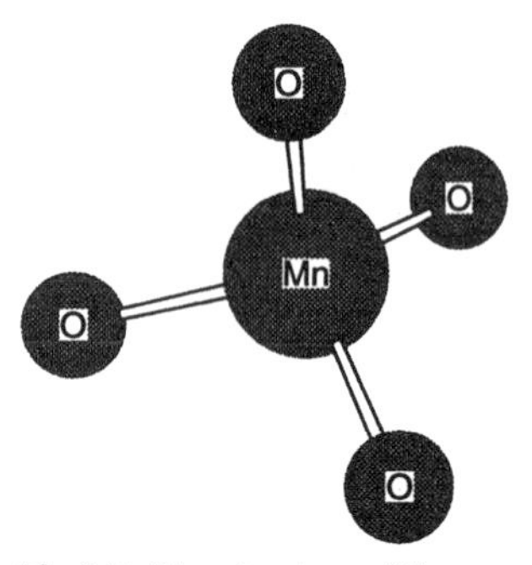

Fig 3.5. The structure of the permanganate anion $[MnO_4]^-$.

Coordination number four is the smallest of the very common coordination numbers. The two most important geometries for four coordinate species are tetrahedral and square planar, however there are many examples of intermediate or distorted geometries. The most common geometry is tetrahedral, as seen in permanganate, $[MnO_4]^-$, (Fig. 3.5). Some other examples of four-coordinate complexes are shown in Fig. 3.6. Tetrahedral coordination is conveniently represented by a sp^3 hybridization model. There is a very important group of d^8 (Chapter 4) metal compounds that are square planar and these are represented by dsp^2 hybridization.

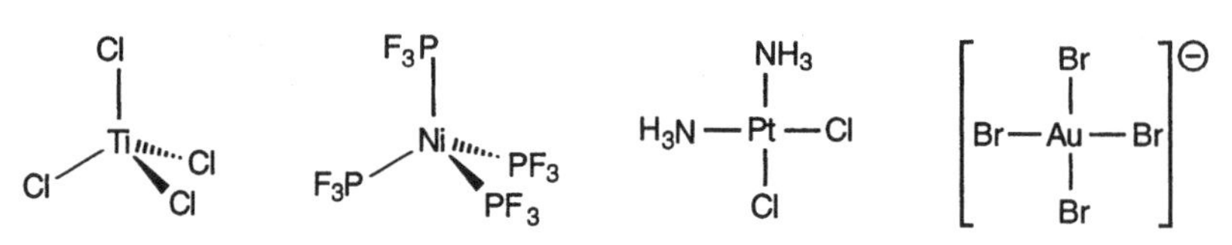

Fig 3.6. Examples of four-coordinate tetrahedral and square planar complexes.

Platinum(II) is a good example of a d^8 metal and there are many examples of square planar Pt(II) complexes. The size of the metal ion is important in determining whether the adopted configuration will be tetrahedral or some higher coordination number. If the metal cation is small, or the ligands are large, then steric effects might compensate for the advantage found in forming more metal-ligand bonds.

d^8 metals include:

Co(I)	Ni(II)	
Rh(I)	Pd(II)	
Ir(I)	Pt(II)	Au(III)

A Rh(I) complex with the empirical formula $RhCl(CO)_2$ has an interesting structure. The empirical formula suggests three coordination but the solid state structure is that shown in Fig. 3.7. The structure is nominally dimeric with two square planar (four-coordinate) Rh(I) ions linked by two bridging chloride groups. However the dimeric units are strung together to form chains in which the dimeric units are connected through rather long Rh···Rh interactions, and if these are included then the overall coordination number is 5. There are examples of more exotic geometries such as the trigonal monopyramidal 2-pyridylmethylbis(2-ethylthioethyl)amine complex in Fig. 3.7 in which the unusual geometry is imposed by the ligand.

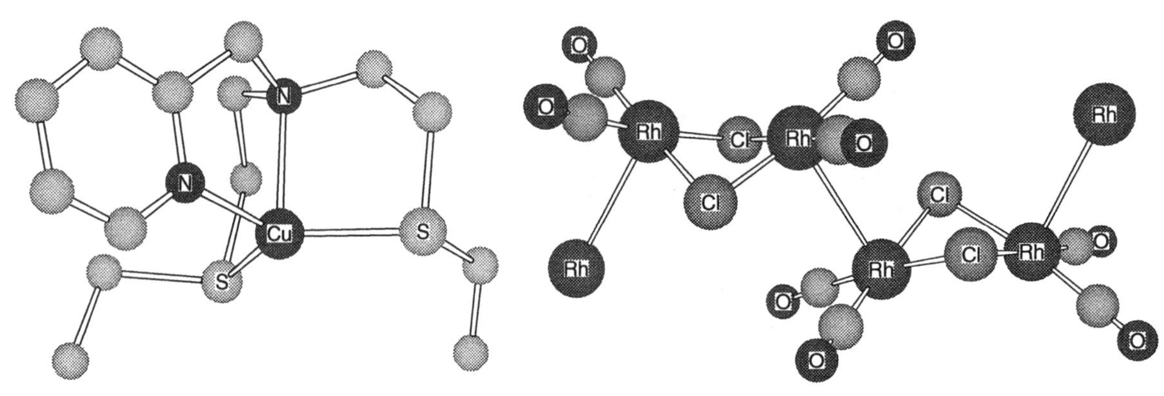

Fig. 3.7. The solid state structure of $[Cu\{N(CH_2CH_2SEt)_2CH_2(C_5H_4N)\}]$ (left) and $[RhCl(CO)_2]_n$.

3.5 Coordination number 5

There are two important five-coordinate geometries of complexes: trigonal bipyramidal and square pyramidal (Fig. 3.8). An analysis of the two geometries suggests that the energy difference between the trigonal bipyramidal and square pyramidal structures is marginal.

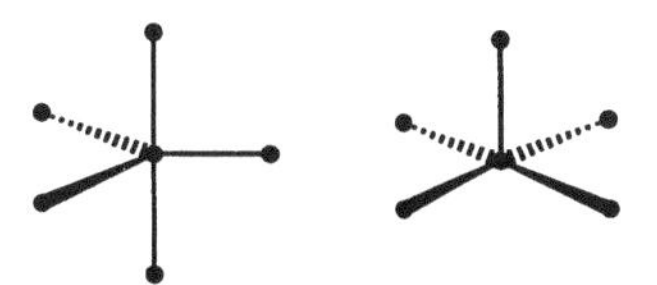

Fig. 3.8. The two main coordination geometries for five-coordination, trigonal bipyramidal (left) and square-based pyramidal.

It is therefore relatively easy to interconvert between these two structures with relatively small movements of the five ligands. Consequently many structures exist with geometries *intermediate* between the two limiting coordination geometries. Further, it is fairly common for the ligands physically to exchange between the sites, in which case the compound is referred to as fluxional.

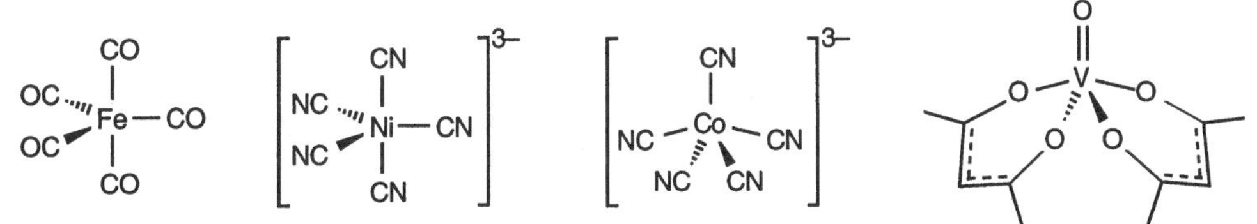

Fig. 3.9. Some examples of five-coordinate complexes.

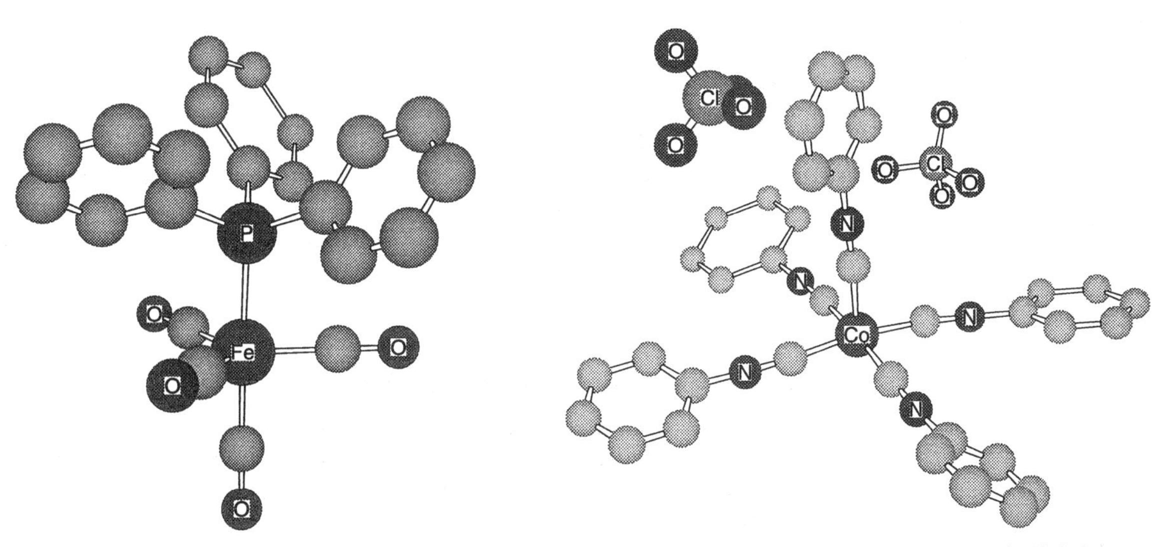

Fig. 3.10. The solid state structures of five-coordinate molecules, $[Fe(CO)_4(PPh_3)]$ (left), and $[Co(CNPh)_5][ClO_4]_2$.

3.6 Coordination number 6

The most common coordination number for *d*-block metal complexes is six. By far, the most common geometry for six coordination is *octahedral*, but the trigonal prismatic geometry is seen occasionally. Some examples of octahedral complexes and a trigonal prismatic complex are shown in Fig. 3.11. Normally the octahedral geometry is fairly regular (Fig. 3.12), but there are many instances of structural distortions brought about by the metal's electronic configuration (Section 7.7). Note that six coordinate complexes with formulae MX_5Y, MX_4Y_2, or MX_3Y_3, etc. are referred to as octahedral

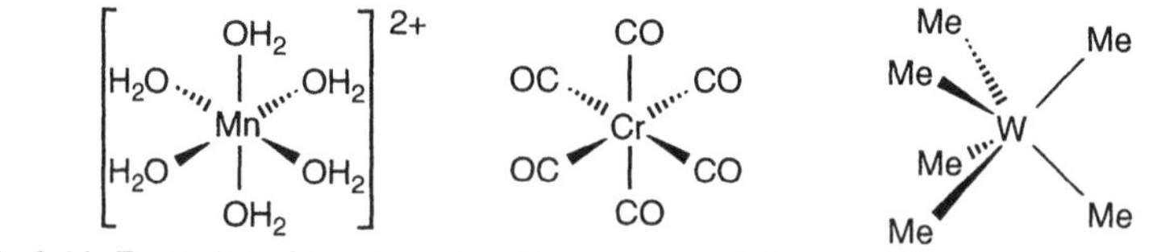

Fig 3.11. Examples of octahedral metal complexes (left, centre) and a trigonal prismatic complex (right).

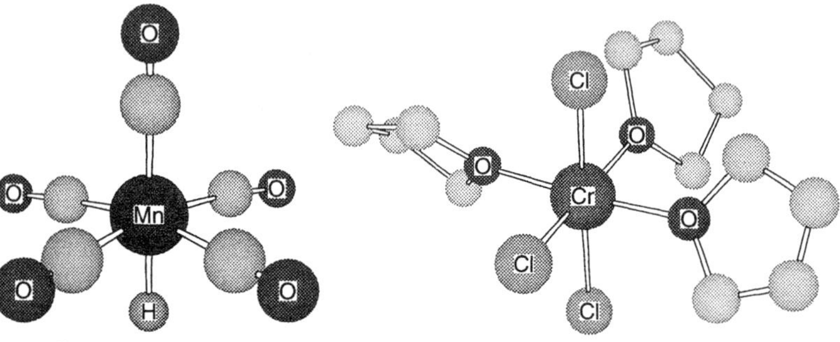

Fig. 3.12. Examples of solid state structures of octahedral metal complexes: $[MnH(CO)_5]$ (left), and $[CrCl_3(THF)_3]$.

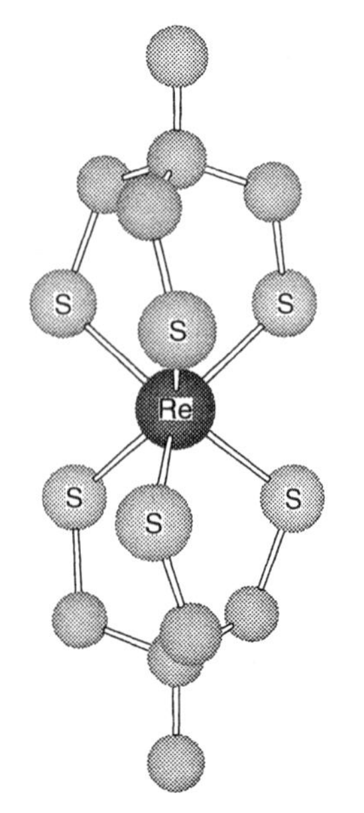

Fig. 3.13. The solid state structure of the trigonal prismatic species $[Re\{S_3(CH_2)_3CMe\}_2]^+$.

complexes even though the ligands may not be exactly at the corners of a regular octahedron.

A few examples of trigonal prismatic geometry are known. In particular, it appears that ML_6 complexes in which the electronic configuration of the metal is d^0 appear to be more stable in the trigonal prismatic geometry than in the octahedral geometry. A good example is the d^0 tungsten complex WMe_6. However, other examples of the trigonal prismatic geometry are known. One example is the rhenium complex $[Re\{S_3(CH_2)_3CMe\}_2]$ (Fig. 3.13).

3.7 Higher coordination numbers

Complexes with coordination numbers higher than six are less common. Ligand–ligand interactions become more important with higher coordination numbers and so it tends to be smaller ligands which allow these coordination numbers. The possibilities for coordination geometries displayed by higher coordination numbers are quite numerous (Table 3.2). Note that the diagrams in Table 3.2 are intended to represent the ideal geometries. The geometries actually displayed are often somewhat distorted, especially if all the ligands are not identical. Thus, there is a series of seven coordinate W(II) complexes described as capped octahedral, but a glance at the structure of one of these, $[WBr_3(CO)_4]^-$ (Fig. 3.14), shows a distinctly distorted structure. The pentagonal bipyramidal geometry is represented by the zirconium fluoride anion $[ZrF_7]^{3-}$ while the anions $[MF_7]^{2-}$ (M = Nb or Ta) both possess the tetragonally capped trigonal prismatic geometry.

Perhaps surprisingly, the cubic geometry is rare amongst discrete (that is, not an extended array) eight-coordinate metal complexes. A few examples are known for the *f*-block elements but it seems that none are known for the *d*-block elements. One neat explanation for this absence is provided by an

Table 3.2. The more common geometries displayed by higher coordination numbers.

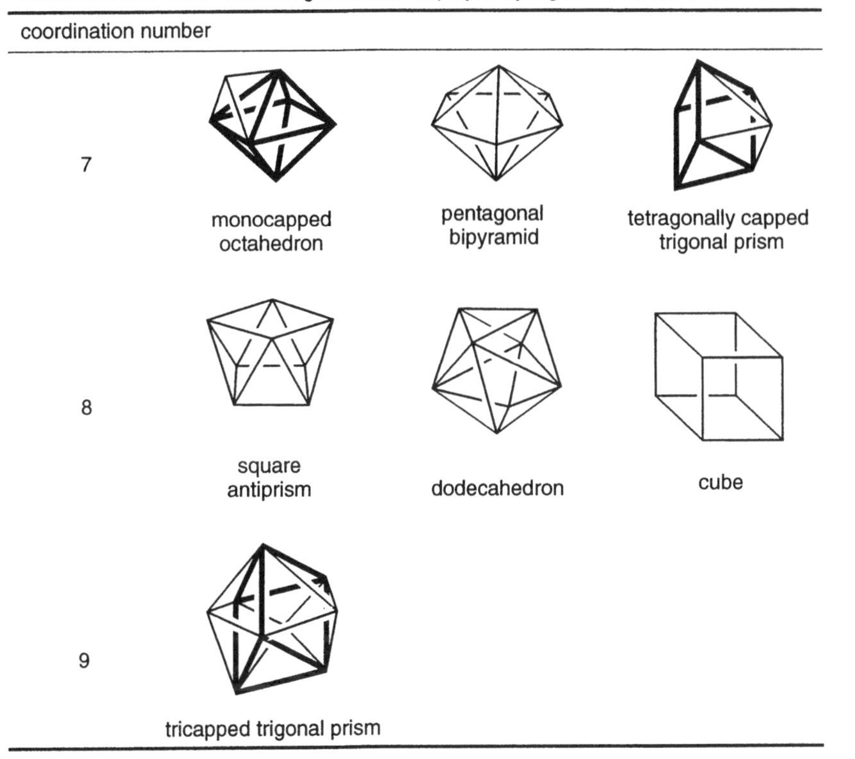

analysis of the symmetry of the cubic ML_8 system. This is discussed very briefly in Section 6.3. The most common geometries for eight coordinate systems are the square antiprism, as in $Na_3[Mo(CN)_8]$, and the dodecahedron as in $(Bu^n{}_4N)_3[Mo(CN)_8]$. Note that the solid state structure is sometimes influenced by the counterion, as in these two cases.

The anion $[ReH_9]^{2-}$ has a tricapped trigonal prismatic structure. Formally, this is a complex of Re(VII) with nine H^- (hydride) ligands. Nine

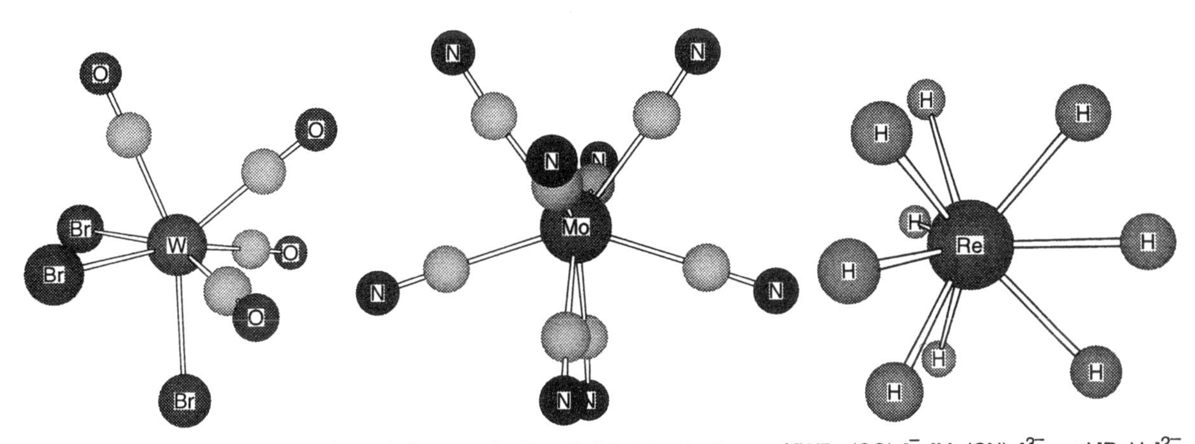

Fig. 3.14. Examples of seven, eight, and nine coordination. Solid state structures of $[WBr_3(CO)_4]^-$, $[Mo(CN)_8]^{3-}$, and $[ReH_9]^{2-}$.

coordination is the limit for covalently bound *d*-block metal complexes since this represents full use of the nine valence orbitals (one *s*, three *p*, and five *d* orbitals).

3.8 Isomerism

Most transition metal compounds are octahedral. There are also many square planar and tetrahedral species. Many other geometries are possible. When two or more complexes of the same empirical formula exist but which have different structures, they are referred to as *isomers*.

There are many ways in which isomers can arise. The number of isomers is dependent upon the coordination number, the stereochemistry of the metal and perhaps upon the nature of the ligand. There are two general classes of isomerism. These are *constitutional isomerism* (the compounds have the same empirical formulae but different atom–atom binding sequences) and *stereoisomerism* (compounds have the same empirical formulae and the same atom–atom sequence, but different arrangements of the atoms in space).

3.9 Constitutional isomerism

Constitutional isomers of compounds have the same empirical formulae, however the atom connectivities differ. There are several different classes of this phenomenon.

Linkage isomerism

Some ligands are able to coordinate to a metal through different ligand atoms. Such ligands are said to be ambidentate. A common example is nitrite, NO_2^-, which in principle can coordinate to a metal as M—NO_2 or M—ONO. The classic example is in cobalt chemistry where the two complexes $[Co(NH_3)_5(NO_2)]^{2+}$ and $[Co(NH_3)_5(ONO)]^{2+}$ are identical except for the coordination of the NO_2^- ligand. The former, N-bonded, species is yellow and the O-bonded species is red. Irradiation with ultraviolet light of the yellow isomer results in interconversion of the yellow nitro complex into the red nitrito complex. The isomerization reverses on warming.

M—NO_2: nitro

M—ONO: nitrito

$$[Co(NH_3)_5(NO_2)]^{2+} \xrightleftharpoons{h\nu} [Co(NH_3)_5(ONO)]^{2+}$$

CNO^-: cyanate

CNS^-: thiocyanate

Other ligands particularly liable to demonstrate linkage isomers are cyanate and thiocyanate. For example, both the square planar complexes $[Pd(NCS)_2(PPh_3)_2]$ and $[Pd(SCN)_2(PPh_3)_2]$ are known. There are many complexes of ambidentate ligands but, of course, in many of those cases not all of the possible linkage isomer complexes are known.

Coordination isomerism

Transition metal complexes may be neutral, cationic, or anionic. In the case of charged complexes, it is quite common to find salts consisting of metal complex cations counterbalanced by metal complex anions so that the salt is electrically neutral. In such cases, isomers might be separable in which the ligands are distributed differently between the two metal centres. For instance the two salts $[Cu(NH_3)_4][PtCl_4]$ and $[Pt(NH_3)_4][CuCl_4]$ are both known and crystallographically characterized. In each case the anions and cations are square planar. Examples are also known for octahedral complexes such as $[Co(NH_3)_6][Cr(CN)_6]$ and $[Cr(NH_3)_6][Co(CN)_6]$.

At this point it is perhaps appropriate to reinforce an earlier warning concerning empirical formulae. There is a cobalt salt $[Co(NH_3)_6][Co(NO_2)_6]$. The empirical formula of this salt is $[Co(NH_3)_3(NO_2)_3]$, which looks like a perfectly reasonable formula, but one which hides the true nature of the complex.

Ionization isomerism

In principle, it might be possible to exchange a ligated anion in a cationic metal complex with the counter anion, which then becomes a ligand in its own right. A good example of this is the pair of compounds $[Co(NH_3)_5Br]SO_4$ (dark violet) and $[Co(NH_3)_5(SO_4)]Br$ (violet–red). These are easy to distinguish by simple chemical tests. The latter reacts with silver ion to precipitate silver bromide, AgBr, whereas the former does *not* give a precipitate with silver ion since the bromide is bonded to cobalt and so not free to react. The tests with Ag^+ ion are complemented by reactions with barium ion, Ba(II). In this case addition of $BaCl_2$ to $[Co(NH_3)_5Br]SO_4$ gives a precipitate of $BaSO_4$ whereas the corresponding reaction of $[Co(NH_3)_5(SO_4)]Br$ does not give a precipitate, as the sulphate ion is bound to the metal and therefore unavailable.

$$[Co(NH_3)_5(SO_4)]Br \xrightarrow{Ag^+} \downarrow AgBr$$

$$[CoBr(NH_3)_5]SO_4 \xrightarrow{Ag^+} \text{no reaction}$$

$$[Co(NH_3)_5(SO_4)]Br \xrightarrow{Ba^{2+}} \text{no reaction}$$

$$[CoBr(NH_3)_5]SO_4 \xrightarrow{Ba^{2+}} \downarrow BaSO_4$$

Solvate isomerism

This is closely related to ionization isomerism and the principle is the same except that a neutral ligand is exchanged for an anionic ligand. One of the best known examples of this involves the complex $[Cr(OH_2)_6]Cl_3$. In this case since the isomerism involves neutral water ligands it is sometimes referred to as *hydrate isomerism*. All three complexes $[Cr(OH_2)_6]Cl_3$, (violet), $[Cr(OH_2)_5Cl]Cl_2.H_2O$, (light green), and $[Cr(OH_2)_4Cl_2]Cl.2H_2O$ (dark green) are known. The last of these is the commercially available material ('Recoura's green chloride') and is derived from a concentrated hydrochloric acid solution. When this salt is dissolved, the complexed chloride ligands are substituted by water, resulting in colour changes from green to light green to violet.

Ligand isomerism

It is quite common for the isomerism phenomenon in complexes to be a consequence of isomerism in the ligand set. At its simplest this might concern complexes of isomeric ligands such as NMe_3 and NH_2Pr.

Sometimes the effect of changing an alkyl chain on a ligand from a linear to a branched structure is sufficient to induce quite profound structural changes in the complex. A good example of this is shown by the reaction of *bis*-(salicylaldehydato)nickel(II) with propylamine to give bis-(*N*-propylsalicylideneaminato)nickel(II).

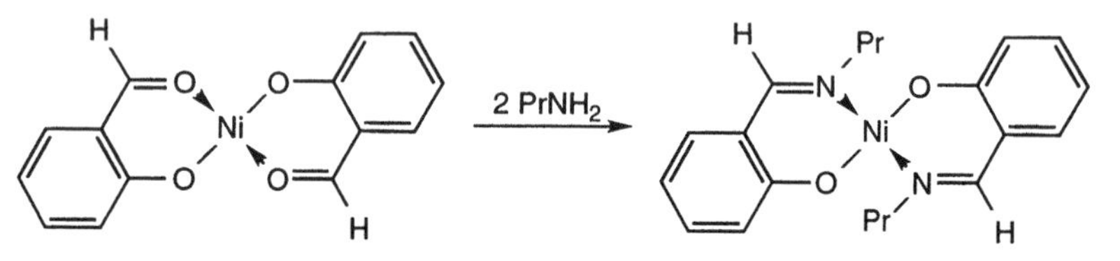

The use of Pr^nNH_2 or Pr^iNH_2 gives two isomers in this reaction. In this case the more sterically demanding Pr^iNH_2 results in a complex distorted away from the square planar geometry displayed by the Pr^n complex towards a tetrahedral geometry. This has consequences for the electronic structure of nickel in these complexes (Chapter 5).

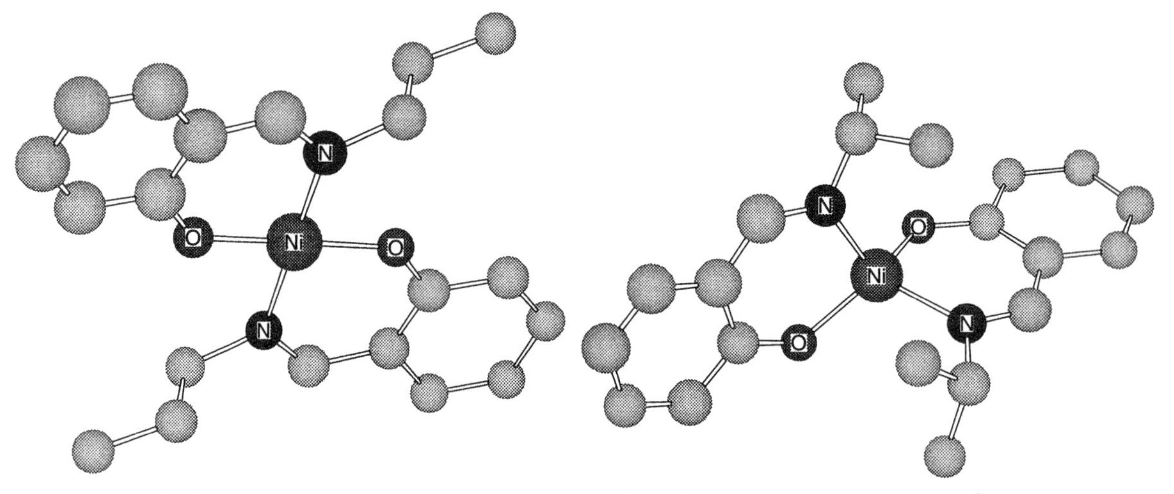

Fig. 3.15. The solid state structures of $[Ni(OC_6H_4CH{=}NPr^n)_2]$ (left) and $Ni(OC_6H_4CH{=}NPr^i)_2]$.

Polymerization isomerism

Polymerization isomerism refers to complexes in which the value of n in the empirical formula $[ML_m]_n$ varies. The pair of complexes $[Pt(NH_3)_4][PtCl_4]$ and $[Pt(NH_3)_2Cl_2]$ possess the empirical formulae $[Pt(NH_3)_2Cl_2]_n$ (n = 1 or 2) and are therefore polymerization isomers. Similarly, the anionic complexes $[Re_2Cl_8]^{2-}$ and $[Re_3Cl_{12}]^{3-}$ might be referred to as polymerization isomers with the empirical formula $[ReCl_4]_n^-$ (n = 2 or 3).

3.10 Stereoisomerism

Stereoisomers possess the same atom–atom connectivities but the individual atoms are arranged differently in space. There are several different types.

Geometrical isomerism

Only one isomer is possible for tetrahedral complexes of formula MA_2B_2 However, Werner discovered *two* isomers for some complexes of general formula MA_2B_2. These compounds are not tetrahedral and are instead square planar, so that two structures are possible. The square planar geometry is an important one and four is the lowest coordination number for which isomerism is important. *Cis* and *trans* square planar complexes provide good examples of geometric isomers.

Diamminedichloroplatinum(II) exists as *cis* and *trans* isomers. These are made by different synthetic routes and exhibit different chemical properties. The *cis*-isomer of $[PtCl_2(NH_3)_2]$ is sold as the drug *cis*-platin and is used for the treatment of, in particular, testicular cancer. The *trans* form is inactive.

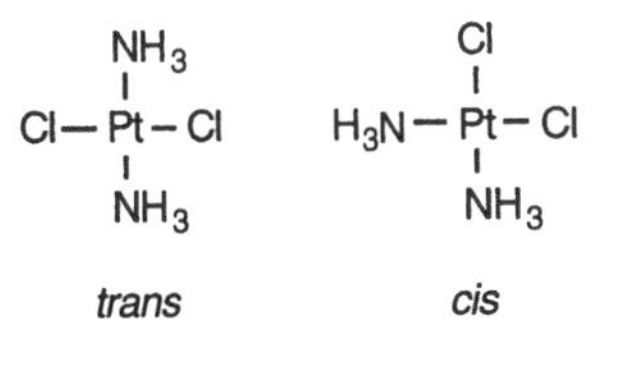

Geometrical isomers are common for six-coordinate compounds. The two chloro ligands in the cation $[Co(en)_2Cl_2]^+$ (Fig. 3.16) may occupy adjacent octahedral positions (*cis* isomer) or opposite positions (*trans* isomer).

In complexes such as $[Co(NH_3)_3Cl_3]$, there are also two ways of arranging the ligands. In the first, two chlorine atoms are mutually *trans* and this is called the *meridional* (*mer*) isomer. In the second, all three chlorine atoms are mutually *cis* and this is the *facial* (*fac*) isomer.

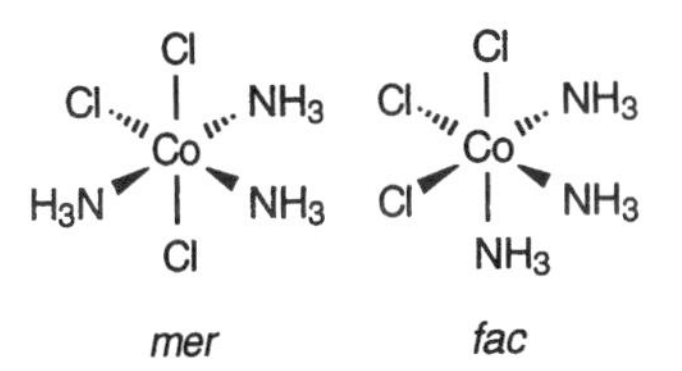

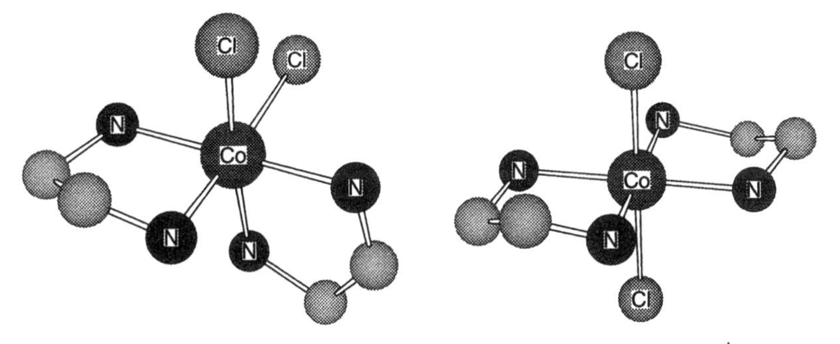

Fig. 3.16. The solid state structures of *cis*- and *trans*- $[CoCl_2(en)_2]^+$.

Polytopal isomerism

This form of isomerism is shown by complexes such as $[Ni(CN)_5]^{3-}$ whose crystal structures show either trigonal bipyramidal or square based pyramidal structures. In fact, the crystal structure of $[Cr(en)_3][Ni(CN_5)]_2.{}^3/_2H_2O$ shows *both* forms at once within the unit cell (Fig. 3.17).

Optical isomerism

A compound that is *not* superimposable on its own mirror image is *chiral*. A pair of distinct isomeric chiral complexes which are mirror images of each other are called optical isomers. The two mirror-image isomers are called enantiomers. One enantiomer rotates the plane of polarized light in one direction whilst the second does so through the same sized angle but in the

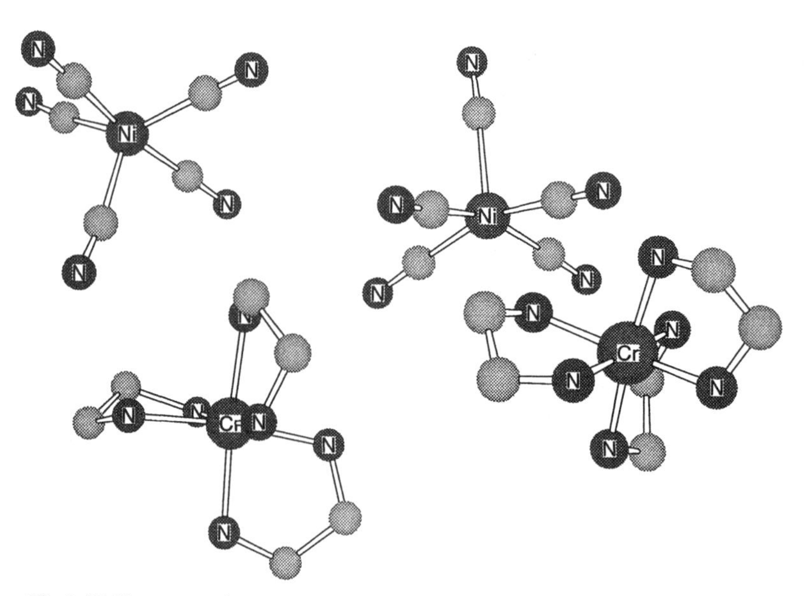

Fig 3.17. X-ray crystal structure of $[Cr(en)_3][Ni(CN)_5]$ (hydrogen atoms and water molecules of crystallization excluded).

opposite direction. Tetrahedral complexes have the potential for optical isomerism just as in carbon derived compounds.

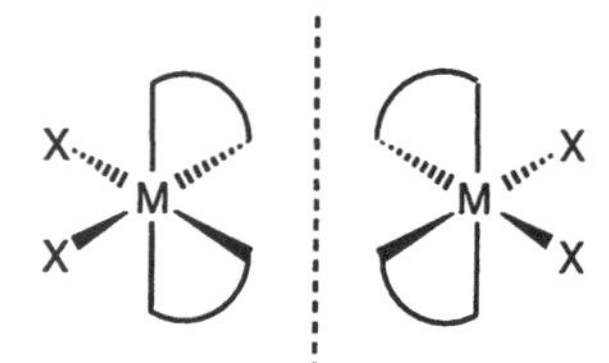

Fig. 3.18. The two non-superimposable mirror images of $MX_2(chelate)_2$.

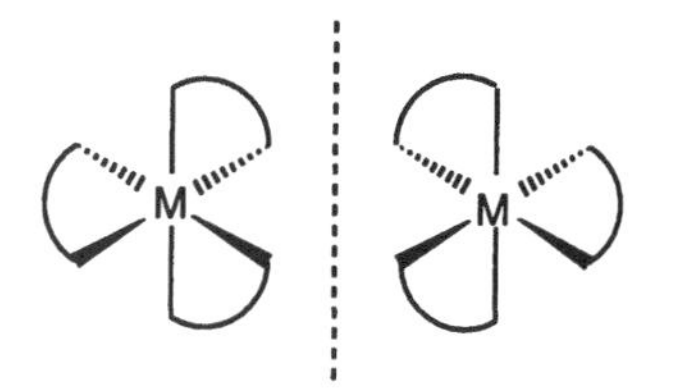

Fig. 3.19. The two non-superimposable mirror images of $M(chelate)_3$.

Optical activity is observed for octahedral complexes containing chelating ligands. The complex cation *cis*-$[CoCl_2(en)_2]^+$ is a good example. The green *trans* isomer has a mirror plane and so is not chiral. The violet *cis* isomer is chiral, exists as nonsuperimposable mirror images, and so is optically active. This is shown schematically in Fig. 3.18 for the general case of an $MX_2(chelate)_2$. Complexes of bidentate chelates with the general formula $M(chelate)_3$ also exist as nonsuperimposable mirror images. The ion $[Cr(en)_3]^{3+}$ (Fig. 3.17) is an example of this and $[Cr(oxalate)_3]^{3-}$ is another. This phenomenon is shown schematically in Fig. 3.19.

One strategy for the resolution of optical isomers is to synthesize the complex with an optically pure counterion, such as tartrate. This leads to a mixture of diastereomers which have different physical properties. These diastereomers may be separated by fractional crystallization as a consequence of their differential solubility, by chromatography, or even by picking out the different crystalline forms as did Pasteur for sodium ammonium tartrate.

3.11 Exercises

1. For each coordination number and each type of isomerism described in this chapter, find one other complex showing that coordination number and isomerism type.
2. How many isomers are possible for a metal complex of the general formula $[MAB(L_2)_2]$ (L_2 represents a didentate ligand such as en). How would this number change if L_2 is changed to a ligand such as $NH_2CH_2CHMeNH_2$?

4 Electron accountancy

In order to understand the chemistry of transition metal compounds, it is necessary to have good models for the bonding in the complexes and to be able to keep track of electrons in those complexes. This chapter sets out some basics of 'electron accountancy' allowing one to keep track of electrons.

4.1 Metal valence electronic configuration

There is often confusion as to the electronic structure of the metals in the ground state and their various oxidation states. The electronic ground states of elements usually quoted are those of *gaseous, neutral atoms*. For the first row transition metals, the $4s$ level fills before the $3d$ level in neutral gaseous atoms. The $3d$ level only starts to fill at Sc ($4s^23d^1$). From Sc the $3d$ shell continues to fill with increasing Z, but note that at Cr there seems to be an anomaly. For chromium, the electrons are distributed so that the 4s and $3d$ levels are each half-filled. The reasons for this configuration's stability are complex and to do with electron–electron repulsions, but beyond the scope of this text. Note that there are a number of further 'anomalous' configurations throughout the d-block elements. These are picked out in bold text in Table 4.1. It is clear that the 'anomalous' configurations make up a considerable proportion of the observed electronic configurations.

However, these s and d level populations are not the configurations found in d-block compounds. When an atom or ion is bound to ligands in a complex, the filling order of the various orbitals reverts to the 'hydrogen

Table 4.1. Valence electron configuration of neutral gas phase d-block atoms.

3	4	5	6	7	8	9	10	11	12
Sc $4s^23d^1$	Ti $4s^23d^2$	V $4s^23d^3$	**Cr $4s^13d^5$**	Mn $4s^23d^5$	Fe $4s^23d^6$	Co $4s^23d^7$	Ni $4s^23d^8$	**Cu $4s^13d^{10}$**	Zn $4s^23d^{10}$
Y $5s^24d^1$	Zr $5s^24d^2$	**Nb $5s^14d^4$**	**Mo $5s^14d^5$**	Tc $5s^24d^5$	**Ru $5s^14d^7$**	**Rh $5s^14d^8$**	**Pd $5s^04d^{10}$**	**Ag $5s^14d^{10}$**	Cd $5s^24d^{10}$
Lu $6s^25d^1$	Hf $6s^25d^2$	Ta $6s^24d^3$	W $6s^25d^4$	Re $6s^24d^5$	Os $6s^25d^6$	Ir $6s^25d^7$	**Pt $6s^15d^9$**	**Au $6s^15d^{10}$**	Hg $6s^25d^{10}$

Table 4.2. Valence electron configuration of complexed neutral d-block metals.

3	4	5	6	7	8	9	10	11	12
Sc $3d^3$	Ti $3d^4$	V $3d^5$	Cr $3d^6$	Mn $3d^7$	Fe $3d^8$	Co $3d^9$	Ni $3d^{10}$	Cu $4s^13d^{10}$	Zn $4s^23d^{10}$
Y $4d^3$	Zr $4d^4$	Nb $4d^5$	Mo $4d^6$	Tc $4d^7$	Ru $4d^8$	Rh $4d^9$	Pd $4d^{10}$	Ag $5s^14d^{10}$	Cd $5s^24d^{10}$
Lu $5d^3$	Hf $5d^4$	Ta $4d^5$	W $5d^6$	Re $4d^7$	Os $5d^8$	Ir $5d^9$	Pt $5d^{10}$	Au $6s^15d^{10}$	Hg $6s^25d^{10}$

order'. That is, in complexes of first row complexes, the $3d$ level is filled *before* the $4s$. Chromium (0) in its complexes therefore has the electronic configuration [Ar]$3d^6$. Similarly, Fe(0) in its complexes is d^8 rather than s^1d^7, V(0) is d^5, while W(0) is d^6, and so on.

4.2 The oxidation state

oxidation: removal of electrons
reduction: acquisition of electrons

When electrons are removed from a metal, M, the metal is said to be *oxidized.* Addition of electrons is referred to as *reduction.* The number of electrons added or subtracted from the neutral metal is the *oxidation state.* The oxidation state is often given in brackets, thus removal of one electron places M in the +1 oxidation state and this is denoted M(I). Addition of two electrons to M places the metal in the –2 oxidation state and this is denoted as M(–II).

A survey of compounds for any one transition metal would show that the metal shows a number of different *oxidation states* within that range of compounds. This willingness to adopt different oxidation states is an important feature of d-block metal chemistry. Metals in positive oxidation states [such as Fe(II)] are referred to as being oxidized from the metal and metals in negative oxidation states [such as Mn(–I)] as reduced.

The electronic configuration of a metal ion in a complex may be determined from the electronic structure of the metal (Table 4.2). Since the d orbitals are the highest lying levels, ionization clearly involves removal of electrons from the d level. Therefore since Os(0) is d^8, Os(II) must be d^6. Since Fe(0) is d^8, Fe(III) must be d^5, and so on. This rule is nearly always valid although there are a few exceptions to the left of the periodic table.

Recall again that the ordering of energy levels in complexes of transition metals reverts to the hydrogen-like order. Therefore electrons contained in the d level are those removed when oxidation occurs and the d level is where electrons are added on reduction. There is therefore a simple relationship between the d-electron configuration, d^n, and the oxidation state (Table 4.3).

In order to discuss the reactions of and bonding in transition metal compounds, a good way of calculating the oxidation state in a metal complex from a knowledge of just the stoichiometry of the compound is required. The oxidation state formalism is useful for electron counting (electron book keeping) purposes. So, how does one determine the oxidation state of a metal in a complex, and therefore the d^n configuration? A simple formalism is adopted. The oxidation state of the metal in a complex is defined as 'the charge remaining on the central metal atom when all the ligands are removed in their *closed shell configuration*'.

A warning: 'Formalisms are convenient fictions which contain a piece of the truth – and it is so sad that people spend a lot of time arguing about the deductions they draw, often ingeniously and artfully, from formalisms, without worrying about their underlying assumptions.' (Roald Hoffmann)

If the ligand is a main group ligand (H_2O, NH_3, NO_2^-, etc.), this means that one formally removes the ligand in such a way that the *octet rule* is satisfied. So a Cl ligand must be removed as Cl^-, since Cl^- has a complete octet of electrons. If O is a ligand, this is removed as O^{2-} since O^{2-} has an octet of electrons. If H_2O is a ligand, it is removed as neutral H_2O, water, since that is the closed shell configuration. Every ligand has a closed shell state. Generally it is fairly clear what these closed shell states are since they are removed in a state which is chemically 'sensible' (it is not expected to

Table 4.3. *d*-Electron configurations of transition metals as a function of oxidation state.

	3	4	5	6	7	8	9	10	11	12
Series 1	Sc	Ti	V	Cr	Mn	Fe	Co	Ni	Cu	Zn
Series 2	Y	Zr	Nb	Mo	Tc	Ru	Rh	Pd	Ag	Cd
Series 3	Lu	Hf	Ta	W	Re	Os	Ir	Pt	Au	Hg
oxidation state										
–IV	7	8	9	10	10	10	10	10	10	10
–III	6	7	8	9	10	10	10	10	10	10
–II	5	6	7	8	9	10	10	10	10	10
–I	4	5	6	7	8	9	10	10	10	10
0	3	4	5	6	7	8	9	10	10	10
I	2	3	4	5	6	7	8	9	10	10
II	1	2	3	4	5	6	7	8	9	10
III	0	1	2	3	4	5	6	7	8	9
IV		0	1	2	3	4	5	6	7	8
V			0	1	2	3	4	5	6	7
VI				0	1	2	3	4	5	6
VII					0	1	2	3	4	5
VIII						0	1	2	3	4

remove a Cl ligand as Cl^+, for instance). As for organic ligands, C_2H_4 comes off as neutral, HC≡CH is also neutral, CO is neutral, benzene (a six electron donor ligand) also comes off as neutral, methyl comes off as Me^-, and so on.

Another definition of oxidation state is that the oxidation state is 'the charge remaining on the central metal atom when each shared electron pair is assigned to the *more electronegative atom*'. Since the transition elements are rather electropositive, these two definitions virtually always lead to the same result. Clearly Cl is more electronegative than all the transition metals and so it is always removed as Cl^-. Note that both H and C are more electronegative than *d*-block metals, so the hydrogen ligand is removed as H^-, and methyl, phenyl, and acyl groups are always removed as Me^-, Ph^-, and $\{C(=O)R\}^-$.

The group H is a perfectly acceptable ligand for a transition metal. The problem here is that one could regard the closed shell configuration for H as either where the 1*s* shell is filled (hydride, H^-) or completely empty (proton, H^+). Since H is more electronegative than most transition metals, it doesn't seem unreasonable to assign both electrons to the H when removing the H ligand. The *convention* or *formalism* is therefore adopted that when the H ligand is bonded to the metal, it is coordinated as hydride, H^-. This is only a formalism and should not be taken to mean that the properties of all transition metal complexes of hydride are hydridic in the sense of LiH, for instance.

The formal oxidation state has little to do with the actual charge on the metal in a complex. It is a formalism that allows one to keep track of electrons. For instance, the true charge of Mn in $[MnO_4]^-$ is not +7, even though the oxidation state is written as Mn(VII). Nor does Pt in $[PtF_6]$ actually exist as the Pt^{6+} ion even though its oxidation state is written as Pt(VI). In fact PtF_6 is a covalent volatile solid which has a conspicuous lack of ionic properties. So, oxidation state does not describe any physical property of a complex. There is no experiment by which it might be measured. Note that the oxidation state has a sign; oxidation states are zero, positive, or negative. Although the oxidation state does not relate to any physical property, it is nevertheless very useful for electronic book keeping purposes and it is essential to be able to assign oxidation states correctly.

The maximum observed oxidation state for a transition metal is normally found for the corresponding fluorides or oxides. This maximum corresponds to the number of valence electrons as far along as Group 7 or 8 but then drops to something lower than the number of valence electrons for elements to the right of the periodic table. The reason for the group number not being exceeded in *d*-block metal chemistry is shown in Fig. 4.1. For each metal the increase in consecutive ionization enthalpy value is fairly steady until a sharp rise in the energy required to remove the next electron is encountered when all the valence electrons have been removed. Thus there is a sharp increase in ionization enthalpy on going from Sc^{3+} to Sc^{4+}, and from Ti^{4+} to Ti^{5+}, and so on. Under normal 'test-tube conditions' once all the valence electrons are removed, it is not possible to break into the inert gas core. Thus, Sc(IV) and Ti(V) are unknown under normal chemical conditions.

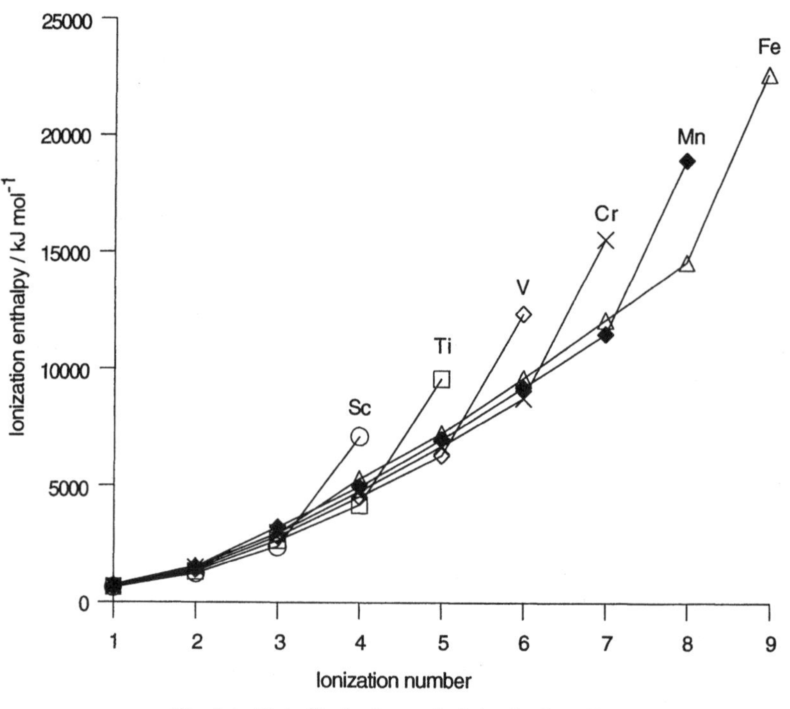

Fig. 4.1. Plot of ionization enthalpies for Sc – Fe.

Examples of oxidation state determinations

The oxidation state of a metal is calculated by determining the difference in the number of valence electrons present in the zero oxidation state (the neutral metal atom) and the number of valence electrons present in the metal

Table 4.4. Examples of oxidation number determination

$[Cr(NH_3)_6]^{3+}$	contribution	running total
number of Cr(O) valence electrons	6	6
charge on complex: +3, assume metal centred,	−3	3
remove six NH_3 ligands, all neutral so no nett effect	0	3
difference between 6 and 3 gives oxidation state		**+3**

$[MnO_4]^-$	contribution	running total
number of Mn(O) valence electrons	7	7
charge on complex: −1, assume metal centred,	+1	8
remove four O^{2-} ligands, therefore subtract 4 x 2	−8	0
difference between 7 and 0 gives oxidation state		**+7**

$[PtF_6]$	contribution	running total
number of Pt(O) valence electrons	10	10
charge on complex: 0, no nett effect	0	10
remove six F^- ligands, therefore subtract 6	−6	4
difference between 10 and 4 gives oxidation state		**+6**

$[Mn(CO)_4]^{3-}$	contribution	running total
number of Mn(O) valence electrons	7	7
charge on complex: −3, assume metal centred,	+3	10
remove four CO ligands, all neutral so no nett effect	0	10
difference between 7 and 10 gives oxidation state		**−3**

$[CoCl_2(en)_2]^+$	contribution	running total
number of Co(O) valence electrons	9	9
charge on complex: +1, assume metal centred,	−1	8
remove two en ligands, all neutral so no net effect	0	8
remove two Cl^- ligands, therefore subtract 2	−2	6
difference between 9 and 6 gives oxidation state		**3**

after removal of the ligands in their closed shell states. If there is any charge on the complex, it is assumed to be localized on the central metal. Thus in $[MnO_4]^-$ (permanganate) start with Mn^-, while in $[Cr(NH_3)_6]^{3+}$ start with Cr^{3+}. Some examples of oxidation determination are shown in Table 4.4.

4.3 Electron counting

Any discussion leading towards an understanding of bonding in metal complexes requires a way of calculating rapidly the total number of valence electrons associated with the metal–ligand bonding system of the complex.

Table 4.5. Examples of total metal valence electron determination.

$[TiCl_4]$	contribution	running total
a Ti(IV) complex, which is d^0	0	0
4 Cl^- 2-electron donor ligands	8	8
total number of valence electrons associated with Ti		**8**

$[CoCl_2(en)_2]^+$	contribution	running total
a Co(III) complex, which is d^6	6	6
2 en 4-electron (2 from each N) donor ligands	8	14
2 Cl^- 2-electron donor ligands	4	18
total number of valence electrons associated with Co		**18**

$[Cu(NH_3)_6]^{2+}$	contribution	running total
a Cu(II) complex, which is d^9	9	9
6 NH_3 2-electron donor ligands	12	21
total number of valence electrons associated with Cu		**21**

$[ReO_4]^-$	contribution	running total
a Re(VII) complex, which is d^0	0	0
4 O^{2-} 4-electron donor ligands	16	16
total number of valence electrons associated with Re		**16**

$[Cr(CO)_5]^{2-}$	contribution	running total
a Cr(-II) complex, which is d^8	8	8
5 CO 2-electron donor ligands	10	18
total number of valence electrons associated with Cr		**18**

The approach is to regard the bonds in a complex as Lewis base–Lewis acid interactions, that is, as a metal atom or ion to which is coordinated a number of ligands through dative coordinate bonds. This is the purpose of having a method of oxidation state determination. A knowledge of the oxidation state means that the number of metal valence electrons is known. All that remains to determine the total number of valence electrons associated with the metal is to add the number of electron pairs donated to the metal from the ligands. Since all the ligands were removed in their closed shell configurations (Cl^-, OH_2, NH_3, OH^-, etc.), all the bonding interactions are dative, each contributing two electrons. Examples of one way to set out these calculations are shown in Table 4.5

4.4 An alternative electron counting method

There is another way to count electrons in complexes. *The final count arrived at is the same by each method.* In the literature, one sees both methods used almost interchangeably. You therefore need to be aware of both methods, and not to get them confused.

This alternative method does not use an oxidation number for the metal. It starts with the number of valence electrons for the *neutral* metal atom. All bonds to the metal are counted *either* as electron pair donation from the ligand to the metal (covalent dative bonding) *or* as shared electron pairs (covalent bonding). Shared electron pairs result in a one electron contribution from the ligand to the metal. In this scheme, methyl, hydride, hydroxide, and halide all count as one electron ligands. Water, ammonia, carbonyl, phosphines, monoalkenes, etc. are all two electron ligands. Oxide counts as double bonded to the metal (M=O) and so is a two electron donor. In all cases, the resulting electron count is necessarily the same as the method outlined earlier.

Table 4.6. Examples of total metal valence electron determination (alternative method).

$MnBr(CO)_5$	contribution	running total
Mn is d^7	7	7
1 Br 1-electron donor ligands	1	8
5 CO 2-electron donor ligands	10	18
total number of valence electrons associated with Cr		**18**

$[V(=O)(SCN)_4]^{2-}$	contribution	running total
V is d^5	5	5
The –2 charge is metal centred	2	2
1 O 2-electron donor ligand	2	9
4 SCN 1-electron donor ligands	4	13
total number of valence electrons associated with V		**13**

4.5 Exercises

1. For each *d*-block metal, attempt to find the formula of complexes for all the oxidation numbers in the range 0–9. You will not find examples of all. Why not? Do you notice patterns of this availability across the periodic table?
2. For each of the complexes you found in question 1, determine the combined number of metal valence and donated ligand electrons. Again, what patterns do you observe?
3. Use the alternative electron counting method outlined in Section 4.4 to determine the total metal valence electron counts for the complexes in Table 4.5, and the 'conventional' method of Section 4.3 to determine the total metal valence electron counts for the complexes in Table 4.6.

5 Bonding: an ionic model

To a first approximation, one can regard *d*-block complexes as either *ionic* or *covalent*. In reality, of course, most compounds lie somewhere in between these two extremes. The bonding in *d*-block metal compounds often can be represented with *either* model and in practice both methods are very useful. For instance, it could well be very reasonable to regard a complex such as $[TiF_6]^{3-}$ as ionic (that is, as a Ti^{3+} ion surrounded by six F^- ions) while it seems a lot less reasonable to think of $[Cr(CO)_6]$, a complex of zerovalent [that is, Cr(0)] chromium with six neutral CO ligands, as ionic. The *crystal field theory* is a useful ionic treatment while *molecular orbital theory* (Chapter 6) provides a good covalent treatment.

One of the most commonly used descriptions for bonding in transition metal compounds is derived from crystal field theory. This bonding model assumes that the bonding between the metal and the ligands in metal complexes is purely *ionic*.

The starting point is to make some *simplifying assumptions*. The metal complex is represented as a *point* positive charge (the central metal ion) surrounded by a set of *point* negative charges (representing the ligand electron pairs). The point positive and point negative charges form the ionic bonding lattice. These point charge approximations seem somewhat drastic, but are nevertheless useful. Bonding energy is produced through *ionic electrostatic forces* between the cation and anions. The bulk of the bonding comes from these electrostatic interactions and is the most important contribution to the bonding in the sense that it is the *largest*. However, this will be largely ignored here since *colour* and *magnetic properties* have relatively little to do with the interionic interactions. Instead, attention is directed to the effect of the ionic environment upon electrons in the *d* orbitals.

5.1 The effect of two *z*-axis electrons on *p* orbitals

First however, it is instructive to examine the effect of ligand electrons on metal *p* orbitals. The reason for this is that in this case the interactions are particularly simple to visualize and because they illustrate an important principle, rather than because of their importance in the bonding of transition metal complexes.

Imagine a set of *p* orbitals associated with an atom lying in free space. Take two electrons and smear them out in a shell so that the electric field is *spherically symmetrical*. If a 'test electron' is placed in any of the *p* orbitals, there will be a coulombic repulsion between this electron and the negatively charged spherical electric field. The effect of this is that the energies of all three *p* orbitals will be *raised* in energy by some amount since the interelectron effect is repulsive. The energy of the three p orbitals is raised *equally*.

Now rearrange the same two equal negative charges from a spherical distribution to a localized distribution along the z axis at equal distances in the positive and negative directions. If a 'test electron' is placed in the p_z orbital, there will be a coulombic repulsion between this electron and the electric field arising from the two negative charges lying along the z axis. The magnitude of this coulombic repulsion is *greater* than the corresponding repulsion between these two negative charges and any electrons placed in either the p_x or p_y orbitals, *simply because electrons in the p_x and p_y orbitals are further away*.

The energies of the field free p orbitals could be taken as a *reference point*, a zero point, against which to make all subsequent measurements. In practice it is often more convenient to take the energy of the p orbitals in the spherical field as the reference point.

Reorganizing the two electric charges so that they lie along the z axis causes the z orbital to be raised in energy from this reference point while the energies of the x and y orbitals drop *below* the zero reference point. The energy of the p_z orbital is raised by twice as much as the energies of the p_x and p_y orbitals are lowered from the spherical field reference point (Fig. 5.1). This is so as to achieve an energy balance. The energies of the p_x and p_y orbital must be equal to each other since they are indistinguishable from each other in every way other than their orientation when in this environment. Further the energies of the p_x and p_y orbitals are *lower* than the energy of the p orbitals in the spherical field. The degeneracy of the original three p orbitals is said to be lifted by the effect of the negative charges lying along the z axis.

The p_x and p_y energy levels are *degenerate* since their energies are precisely equal and required to be precisely equal by the symmetry of the situation

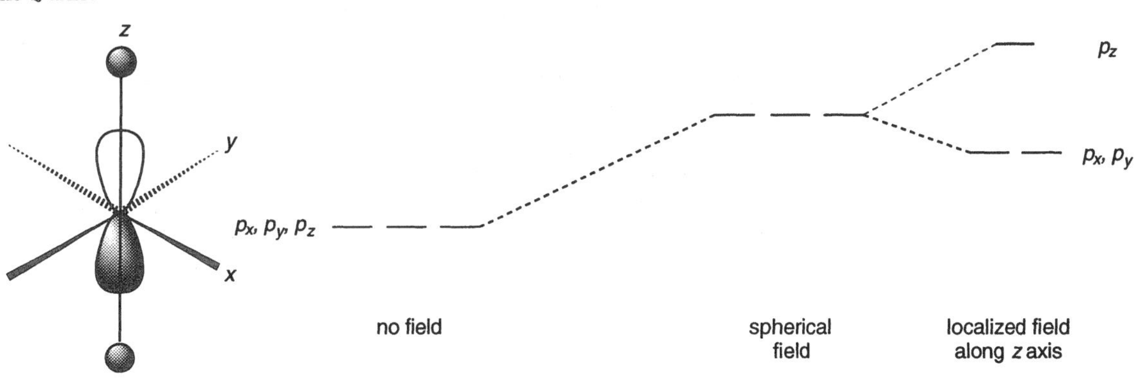

Fig. 5.1. The effect of z-axis electrons upon the energies of p orbital electrons.

5.2 The effect of point charges situated on the coordinate axes upon *d* orbitals

Imagine a metal cation in free space. All the five d orbitals (Fig 5.2) have the same energy (degenerate) and this is represented by the five lines lying at the same level on the left in Fig. 5.3. Now place this ion in a spherically symmetrical electric field consisting of six negative charges. Electrons contained in the metal d orbitals would interact repulsively with this electric

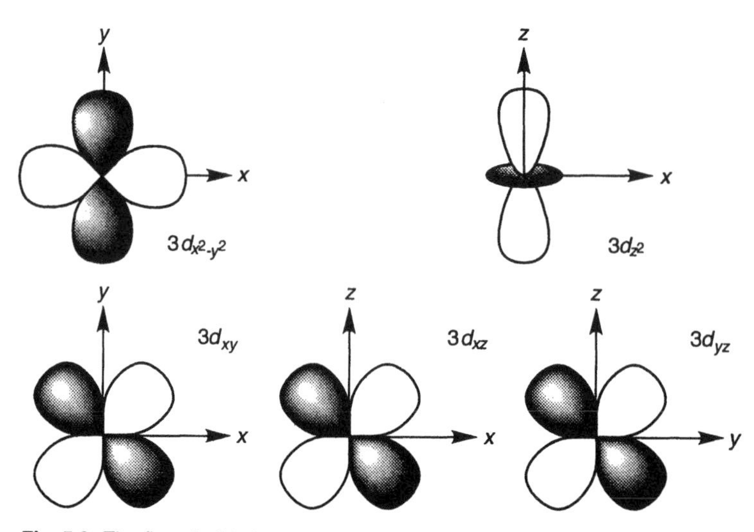

Fig. 5.2. The five *d* orbitals. Note the changes in axis labelling for each orbital.

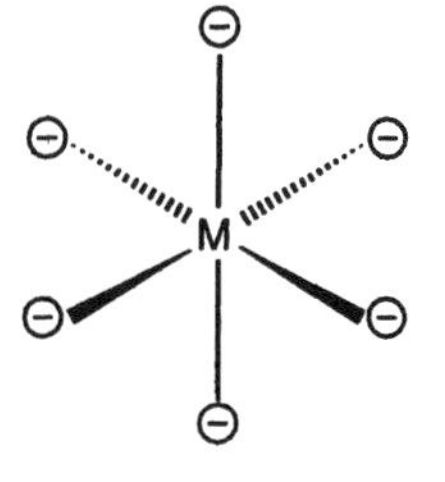

field and this would have the effect of raising the energy of the *d* orbitals equally by some amount as shown.

Now manipulate the electric field so that instead of being a spherically symmetrical field generated by six charges, it is a field generated by placing six point charges at the vertices of an octahedron on the *x*, *y*, and *z* axes. The total charge is the same but its distribution is different. As in the *p* orbital case above, the energy of the five *d* orbitals in a spherically symmetrical field is adopted as the reference level energy. Inspection of Fig 5.2 shows that the $d_{x^2-y^2}$ and d_{z^2} orbitals are those that point most directly towards the six point charges representing the ligands. These are therefore the two orbitals which interact most strongly with the ligand charges and whose energies are raised the most (Fig. 5.4). The other three orbitals, the d_{xy}, d_{xz}, and d_{yz} orbitals energies are not affected so much. In fact, relative to the energy of the five *d* orbitals in a spherically symmetrical field, their energy is lowered.

Note that the energies of the d_{z^2} the $d_{x^2-y^2}$ orbitals are precisely equal, that is, these two orbitals are degenerate. Similarly the d_{xz}, d_{xy}, and d_{yz} orbitals are also degenerate. Inspection of Fig. 5.2 makes it clear that the d_{xz}, d_{xy}, and d_{yz} orbitals are in equivalent environments in the octahedral geometry but the

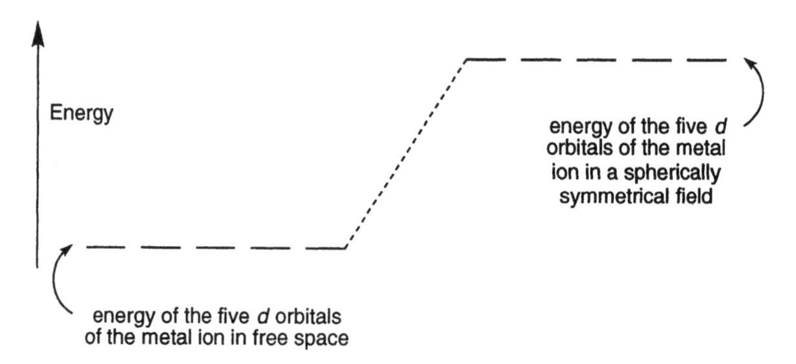

Fig. 5.3. The effect of a spherical field totalling six charges on the metal *d* orbitals.

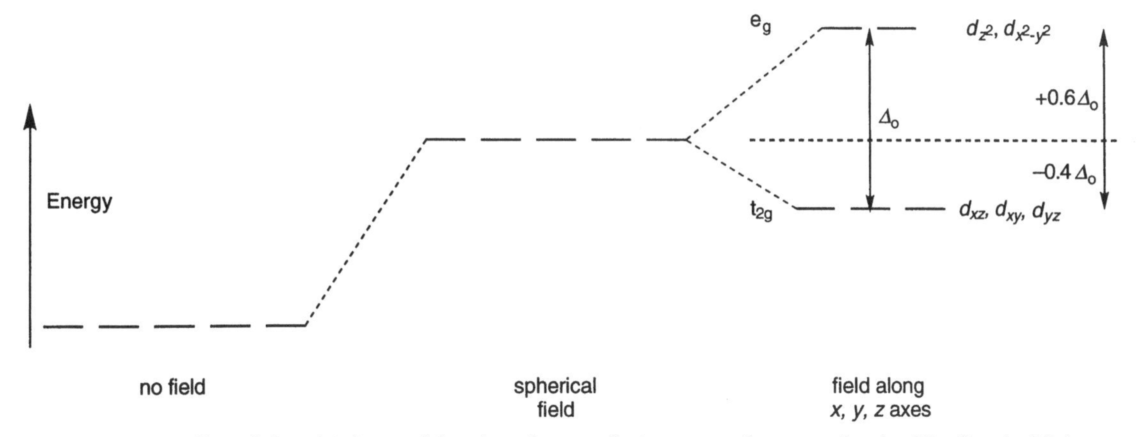

Fig. 5.4. The effect of six point charges lying along the coordinate axes on the energy levels of the five *d* orbitals.

reasons why the d_{z^2} and $d_{x^2-y^2}$ orbitals should be so related are less clear. However, mathematically, the d_{z^2} orbital is a linear combination of $d_{z^2-x^2}$ and $d_{z^2-y^2}$ orbital functions (that is, orbital functions with the same shape as the $d_{x^2-y^2}$ orbital, but pointing in different directions). Once this is appreciated, then it is not unreasonable to accept that the d_{z^2} orbital is affected by exactly the same amount as the $d_{x^2-y^2}$ orbital.

The pattern of one doubly and one triply degenerate orbital sets in Fig. 5.4 is characteristic of all octahedral complexes. The set of three degenerate orbitals is given a name: t_{2g}. The origins of this name lie within a branch of mathematics called group theory and it is beyond the scope of this text to examine these origins. The doublet is called the e_g level and again the origins of this name are from group theory.

Δ_o: the crystal field splitting parameter

The two levels t_{2g} and e_g are split in energy. The difference is labelled Δ_o which is called the *crystal field splitting parameter*. The subscript 'o' indicates that an octahedral complex is under discussion. The actual value of Δ_o depends on the particular compound and varies with the metal, the ligands, and the charge on the complex. Some of these factors will be examined later.

As for the *p* orbital case, an *energy balance* is struck between the energies of the *d* orbitals in the spherical field. There are three orbitals whose energy is falling and two whose energy is rising. Therefore if the t_{2g} set drops by *two* units a balance is achieved if the e_g set rises by *three* units. If the difference between the t_{2g} and e_g levels is defined as Δ_o, then it is clear that the energy of the t_{2g} level changes by $-0.6\Delta_o$ while the energy of the e_g set changes by $+0.4\Delta_o$, all relative to the spherical field reference point.

$10Dq = \Delta_o$

The unit Δ_o is in common usage. Some texts use Dq units instead of Δ_o units. The conversion is simple since 10Dq is equivalent to Δ_o. In the octahedral case, this means that the t_{2g} level is stabilized by 6Dq while the e_g level is destabilized by 4Dq relative to the spherical field reference point. The unit Δ_o is preferred since it explicitly states the geometry of the complex and so avoids confusion when discussing other geometries for which a different value of the crystal field splitting energy is used.

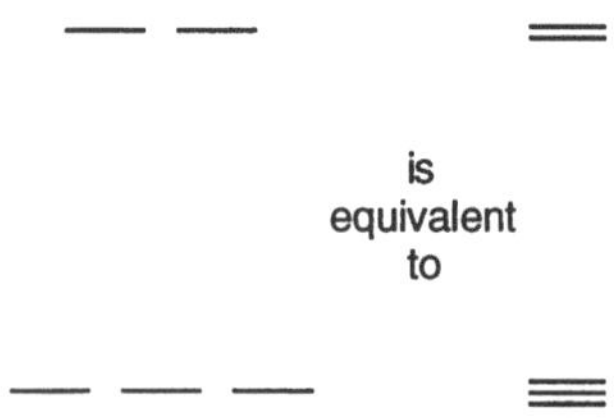

is
equivalent
to

Fig. 5.5. The stacking convention for energy level diagrams.

It is immediately clear that the doublet e_g and triplet t_{2g} levels are degenerate when displayed alongside each other, as in Fig. 5.5. However this way of writing the levels does take up quite a bit of room and diagrams for more complicated systems can look rather untidy. For purposes of conserving space a common convention is employed in many texts for displaying energy level diagrams. The stacking convention is not technically correct since it could be taken to imply sets of non-degenerate orbitals with very slightly different energies, but is nevertheless in common use. In some books sets of degenerate levels are written as just a single line to represent two or three orbitals. This is far less desirable and requires that the reader is conversant with *group theory* terminology. In such cases the reader must remember that whenever the name of the level on the diagram starts with an e it means that the energy level is doubly degenerate and whenever it starts with a t, it means that the level is triply degenerate.

5.3 Crystal field splitting for a cubic ML_8 complex

The cubic ML_8 system is examined here, not because it is common, but because it is an intuitively useful way to approach an analysis of *d* orbital splitting in tetrahedral systems, which *are* common.

As for the octahedral field, the approach is to start from a spherically symmetrical field, although this time forming *eight* point charges. The effect of the spherical field upon the five *d* orbitals is to raise them in energy. Clearly this is very similar to that of the spherical field involving six charges discussed for the octahedral case, but a little more so. The eight charges are then localized so that they lie on the eight corners of a cube.

A simple geometrical argument shows that the $d_{x^2-y^2}$ orbital is *less* strongly affected by the eight ligands than the d_{xy} orbital (Fig. 5.6). That is, an electron in the $d_{x^2-y^2}$ orbital is effectively further away from the eight charges representing the ligands than an electron in the d_{xy} orbital. The $d_{x^2-y^2}$ orbital will therefore be lower in energy than the d_{xy} orbital. By symmetry, the d_{xz} and d_{yz} orbitals are affected by exactly the same amounts as the d_{xy} orbital. For the same reasons alluded to in the previous Section, the d_{z^2} orbital is affected by exactly the same amount as the $d_{x^2-y^2}$ orbital.

Compare the effects of the eight charges upon the $d_{x^2-y^2}$ and d_{xy} orbitals.

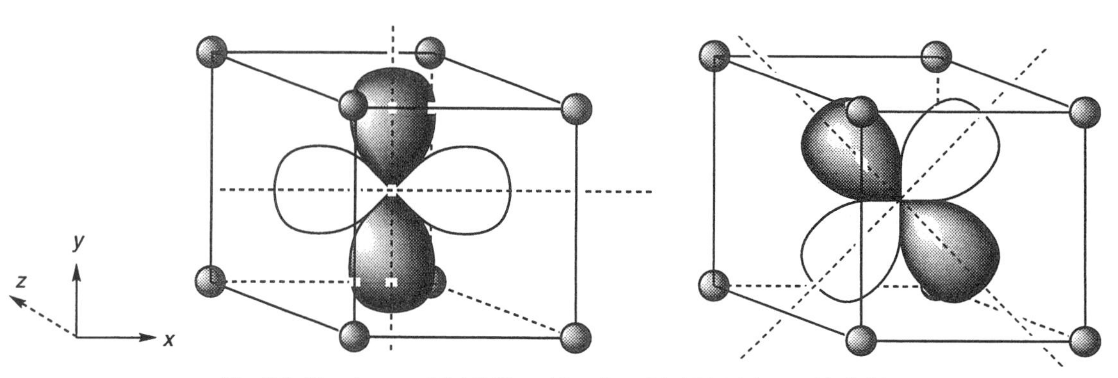

Fig. 5.6. The $d_{x^2-y^2}$ orbital (left) and the d_{xy} orbital (right) in a cubic field.

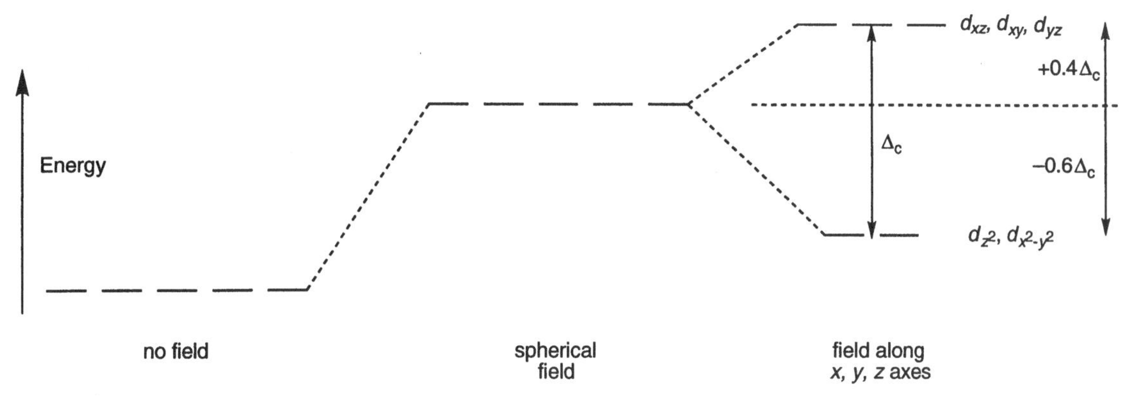

Fig. 5.7. *d* Orbital splitting in a cubic field. The gap between the energy levels is called Δ_c, the c standing for cubic.

The $d_{x^2-y^2}$ orbital is affected *less* by the eight ligands forming the cubic field than the d_{xy} orbital. Its lobes point less directly at the charges than those of the d_{xy} orbital. Effectively a $d_{x^2-y^2}$ electron is further away from the eight point charges than a d_{xy} electron. This will mean that this orbital is stabilized relative to the d_{xy} orbital.

The resulting energy level diagram is shown in Fig. 5.7. Notice that in this case, the ordering of the *d* orbitals is *reversed* from the ordering in the octahedral case. The difference in energy between the sets of levels is called Δ_c, the c standing for cubic.

5.4 Crystal field splitting for a tetrahedral ML_4 complex

The same basic procedure is applicable for the tetrahedral field as for the octahedral geometry. The only problem is that it is less clear intuitively which orbitals are most affected when four ligands are arranged along tetrahedral axes. The way forward is to realize that four opposite corners of a cube define a tetrahedron (Fig. 5.8). A tetrahedral complex is derived conceptually from a cubic complex by removal of four ligands from opposite corners as shown. The four ligands that are removed themselves define a tetrahedron of the same size as the tetrahedron remaining.

The cartesian axes are defined as those passing through the centre of the

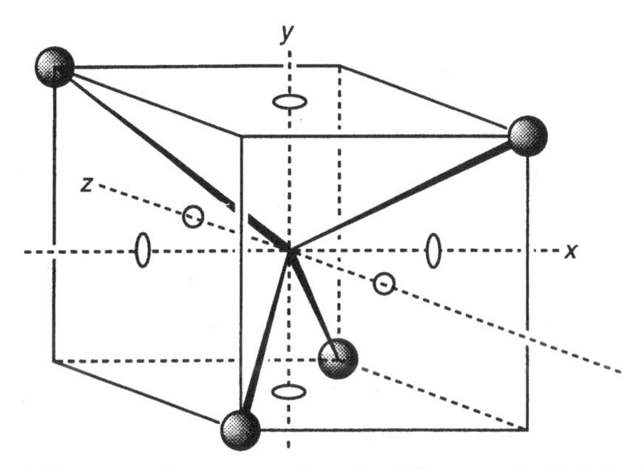

Fig. 5.8. Four opposite corners of a cube define a perfect tetrahedron.

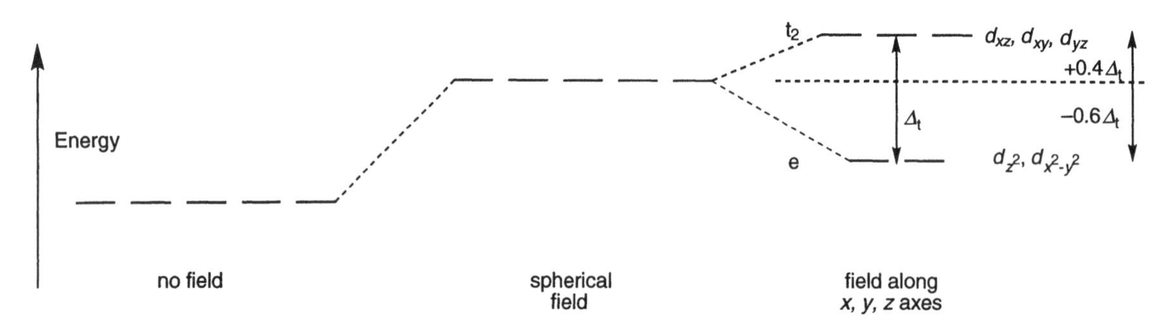

Fig. 5.9. *d* Orbital splitting in a tetrahedral field. The gap between the energy levels is called Δ_t, the t standing for tetrahedral.

cube and the centre of each cube face. The situation is really nearly identical to that of the cubic field. The $d_{x^2-y^2}$ and d_{z^2} orbitals (labelled e) are stabilized relative to the d_{xy}, d_{xz}, and d_{yz} orbitals (labelled t_2), but only by about *half* as much as in the cubic case (Fig. 5.9). It is not unreasonable that the splitting is about half the size of the splitting in the corresponding ML_8 complex (Fig. 5.10). After all, there are exactly half the number of orbitals. However, Δ_t is only about half the value of Δ_c, not exactly half since the M—L bond lengths in ML_4 and ML_8 complexes need not be identical (one factor that affects values of Δ is the M—L bond length).

$\Delta_t = {}^4/_9\Delta_o$

Assuming constant bond lengths, a simple geometric argument shows that Δ_t is smaller than Δ_o and that the relationship between them is $\Delta_t = {}^4/_9\Delta_o$.

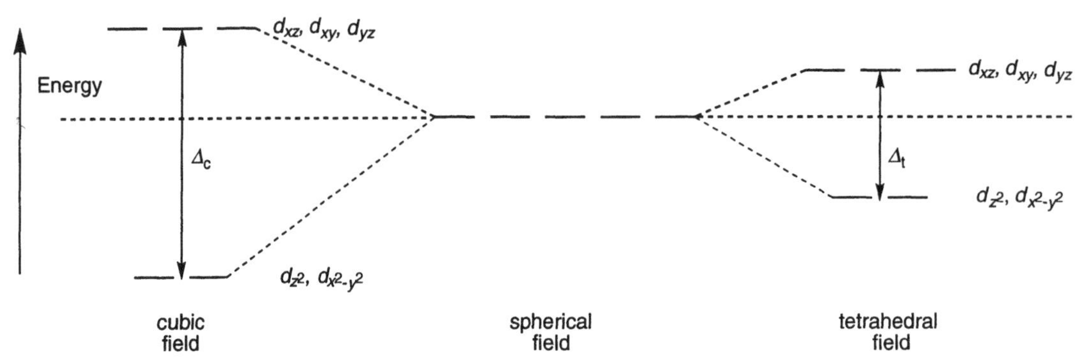

Fig. 5.10. The relationship between *d* orbital splitting in a tetrahedral and a cubic field.

5.5 Crystal field splitting for a square planar ML_4 complex

At this point it is useful to consider square planar complexes. The square planar geometry is another common geometry displayed by certain types of complex, particularly metals in the d^8 configuration.

When building up the energy level diagram for octahedral and tetrahedral complexes, the approach was to place a spherical electric field around the metal and then to organize it into point charges to represent the set of ligands. One could follow exactly the same approach for the square planar system, but it is instructive to look at this system in a slightly different fashion. One way to generate conceptually a square planar set of ligands is to move two of the mutually *trans* ligands in an octahedral complex out to infinity.

Since the octahedron is symmetrical, it doesn't matter which two mutually *trans* point charges are removed, but the axis from which those charges are removed is *defined* as the z direction. A partial removal of ligands along the z direction gives a distorted octahedron. This distortion is called a tetragonal distortion since there is fourfold symmetry. The effect of movement of two ligands out along the z axis is to stabilize the orbitals which are directed either directly or partially along the z direction.

Consider first the e_g set. The e_g set consists of the $d_{x^2-y^2}$ and d_{z^2} orbitals. The effect of starting the move of the z axis charges is that the d_{z^2} orbital clearly does not interact quite as strongly with those charges as they move away. The energy of the d_{z^2} drops a little (Fig. 5.11): there is less repulsion between an electron in the d_{z^2} orbital and the z axis charges. Now consider the effects of this partial removal of the z axis charges on the t_{2g} set. The effects here are less marked since the d_{xy}, d_{xz}, and d_{yz} orbitals do not point so directly towards the coordinate axes. The consequence is that orbitals with a component in the z direction are stabilized (the d_{xz} and d_{yz} orbitals).

The next logical step is to effect the complete removal of these two charges (remove them to infinity). The resulting diagram shows the arrangement of the five d orbitals in a square planar field.

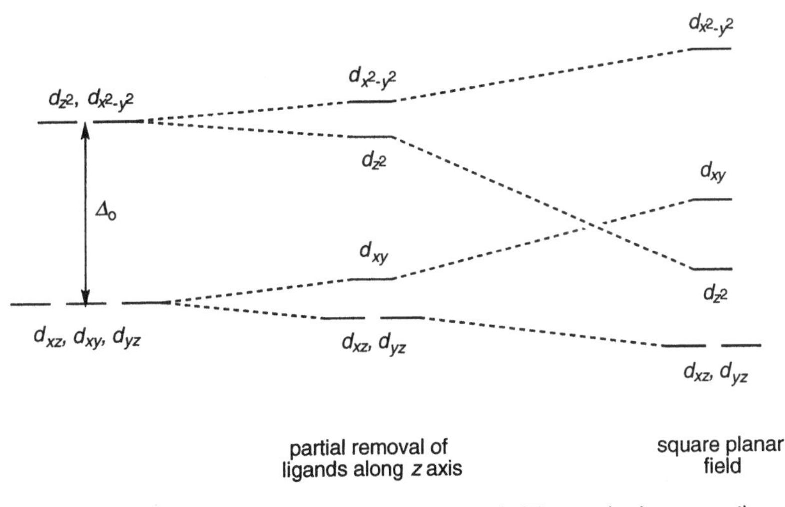

Fig. 5.11. The effect of partial and complete removal of the z axis charges on the octahedral crystal field.

5.6 The d^n configurations of octahedral complexes

Having seen how to determine the number of metal valence electrons in a complex, the next thing is to address in which orbitals the electrons reside when in a metal complex. Remember the crystal field model is ionic. In setting up the ionic model (formation of metal cation) the *number* of d electrons in the cation was fixed. It is now necessary to arrange these electrons in the d orbital t_{2g} and e_g energy levels.

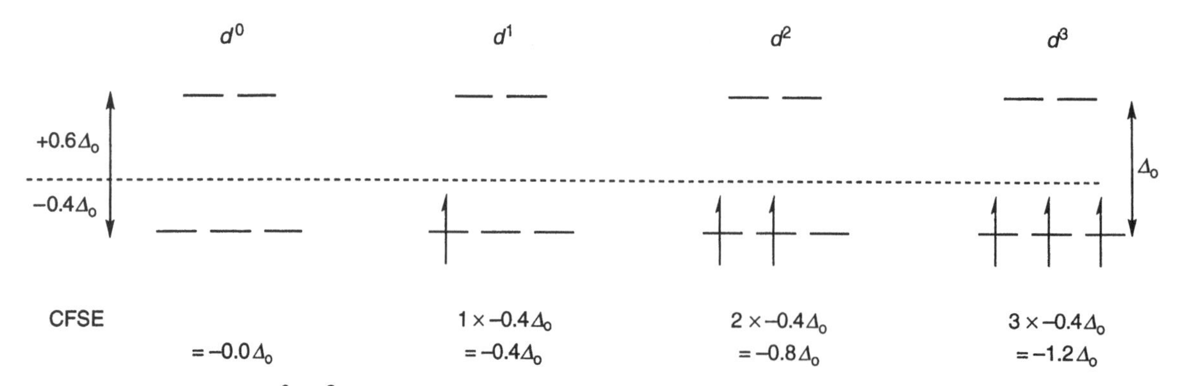

Fig. 5.12. The $d^0 - d^3$ configurations for octahedral complexes. Note the spin parallel configurations.

The $d^0 - d^3$ configurations

The configurations for the $d^0 - d^3$ cases are shown in Fig. 5.12. If the electronic configuration of the metal is d^0, then there will be no electrons in the d orbitals in the complex. In the d^1 case, the electron occupies the lowest available orbital. This is one of the t_{2g} set, for now, it does not matter which. The electron is $0.4\Delta_o$ lower in energy relative to the energy of the five d orbitals in the electric field (the dotted line). Since positive values go upwards, and since the dotted line is the reference point, the d^1 configuration is stabilized by $-0.4\Delta_o$ relative to the reference point. This stabilization is called the *crystal field stabilization energy* (CFSE).

In the d^2 configuration there are two electrons, each of which is stabilized by $-0.4\Delta_o$ and so the total stabilization is $2 \times -0.4\Delta_o = -0.8\Delta_o$. There are two other points to note about the electronic configuration. First, the second electron does not occupy the same orbital as the first electron. There would be an energy penalty in so doing. Second, the orientation of the spin is parallel to that of the first. This follows Hund's rules.

For the d^3 configuration there are three electrons to put into the t_{2g} set. These go one into each orbital with their spins parallel. The CFSE is therefore given by $3 \times -0.4\Delta_o = -1.2\Delta_o$.

The CFSE often is not quoted in kJ mol^{-1} or some other absolute energy unit. Instead, the *variable unit* Δ_o is often more convenient. The value of Δ_o varies according to the complex but it is a measurable quantity. While for, say, a pair of d^2 complexes the values of Δ_o expressed in kJ mol^{-1} will probably be different, the CFSE is always $-0.8\Delta_o$, whatever that might mean in absolute terms.

The $d^4 - d^7$ configurations

At the d^4 configuration [Cr(II), perhaps], there is a problem. Where is the fourth electron to go? Consider the energy levels (Fig. 5.12) for the d^3 configuration. If a fourth electron were to go into the t_{2g} set, it would have to spin pair with an electron in one of the orbitals, since they are now all half-filled. There is an energy penalty to pay for pairing up the two electrons. This is called the pairing energy. On the other hand, if the electron were to go into the e_g level, extra energy is required to push it up over the dotted line, the

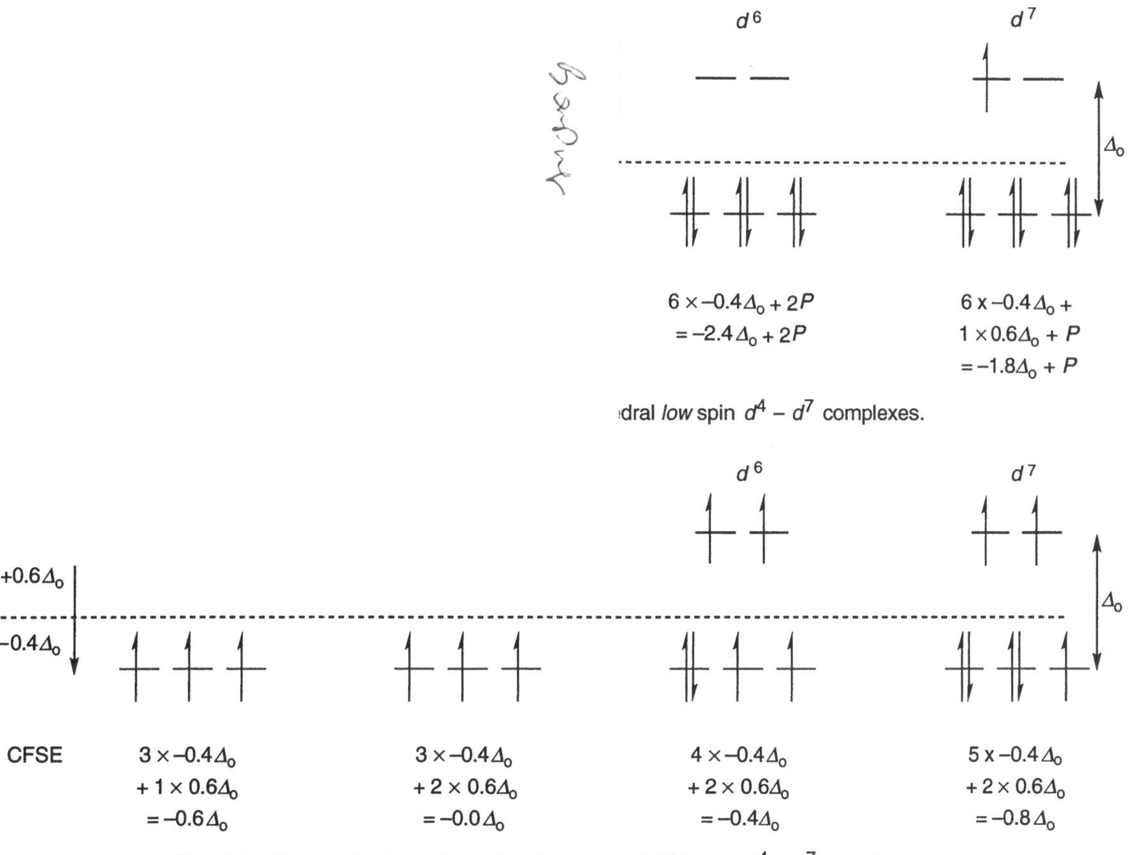

dral *low* spin $d^4 - d^7$ complexes.

Fig. 5.14. The electronic configuration of octahedral *high* spin $d^4 - d^7$ complexes.

reference point in energy terms. These two situations are shown in Figs 5.13 and 5.14. In the first case, the contribution to the CFSE for the fourth electron is given by $-0.4\Delta_o + P$ (where P is the pairing energy). In the second case, the contribution to the CFSE for the fourth electron is given by $+0.6\Delta_o$ (a destabilization). In practice, both situations arise, and which configuration is actually adopted depends on the value of Δ_o.

In the first case, there are *two* unpaired electrons in the resulting energy level diagram. In the second case there are *four* unpaired electrons. Since there are fewer unpaired electrons in the first configuration than in the second, where there are four, the first configuration is called *low spin* and the second is called *high spin*.

The CFSE calculations for the d^4 configurations are shown in Figs 5.13 and 5.14. In the low spin case the overall CFSE is $-1.6\Delta_o + P$ and in the high spin case the CFSE is $-0.6\Delta_o$. Given that the fourth electron has to go somewhere, it is clear that if P is larger that Δ_o, then the complex would be better off adopting the high spin configuration, whereas if P is smaller than Δ_o, it is better off in the low spin configuration. This is shown in Fig. 5.15. The centre block of five energy levels represents the five degenerate d orbitals in a spherically symmetrical field. The situation on the left is the low spin case for which Δ_o is greater than the pairing energy. The situation on the

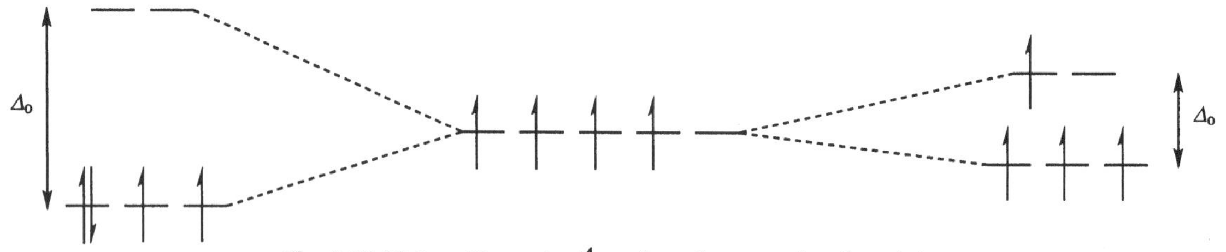

Fig. 5.15. High and low spin d^4 configurations as a function of Δ_o.

right is the high spin case for which Δ_o is less than the pairing energy. For the d^4 case in the spherically symmetrical field, there are no paired electrons. In the octahedral field in the low spin case two of the electrons are paired, hence the value P must be included in the CFSE calculation. In the high spin case there are no unpaired electrons so no value for P has to be included.

If the fourth electron of the five has gone into the t_{2g} level (low spin), this is because the pairing energy is smaller than Δ_o. The pairing energy is still less than Δ_o for the fifth electron, since nothing has changed, and so the fifth electron will *also* go into the t_{2g} set. On the other hand, if the fourth electron of the five has gone into the e_g level (high spin), this is because the pairing energy is greater than Δ_o. The pairing energy is still greater than Δ_o for the fifth electron, since nothing has changed, and so the fifth electron will *also* go into the e_g set. What cannot happen is for the fourth electron to go into the t_{2g} set and the fifth to go into the e_g set. In other words the configuration $t_{2g}^4e_g^1$ is illegal as a ground state configuration – but it could be a spectroscopic *excited state.* There are no unpaired electrons in the d^5 symmetrical field, all five orbitals have just the one electron. In the d^5 low spin octahedral field, there are two pairs of electrons, hence a value of $2P$ is required in the CFSE calculation. There are no paired electrons in the high spin case, so no value for P is required in the CFSE calculation.

Another important point is evident for the d^6 case (Fig. 5.12 and 5.13). The CFSE for the d^6 state only contains a $2P$ quantity rather than $3P$ although there are three paired electrons in the t_{2g} state. Since there are six electrons and five orbitals, pairing of two electrons is required for the d^6 configuration in the symmetrical field. There are three pairs of electrons in the low spin case. Since the CFSE calculations give stabilization energies relative to the symmetrical field, although there are three pairs, there are only *two* more than already present in the symmetrical field. This is why the CFSE calculation shows a value of just $2P$ rather than $3P$ for the low spin d^6 case. Since there is the same number of electron pairs in the d^6 high spin case as for the symmetrical field, no allowance for electron pairs needs to be made.

There are two pairs of electrons for the d^7 configuration in the symmetrical field, and three in the low spin configuration. There is therefore only one extra pairing in the low spin d^7 case relative to the d^7 configuration in a spherical field. Therefore only one P appears in the calculation. Again, since there is the same number of electron pairs in the d^7 high spin case as for the symmetrical field, no allowance for electron pairs needs to be made for the high spin case.

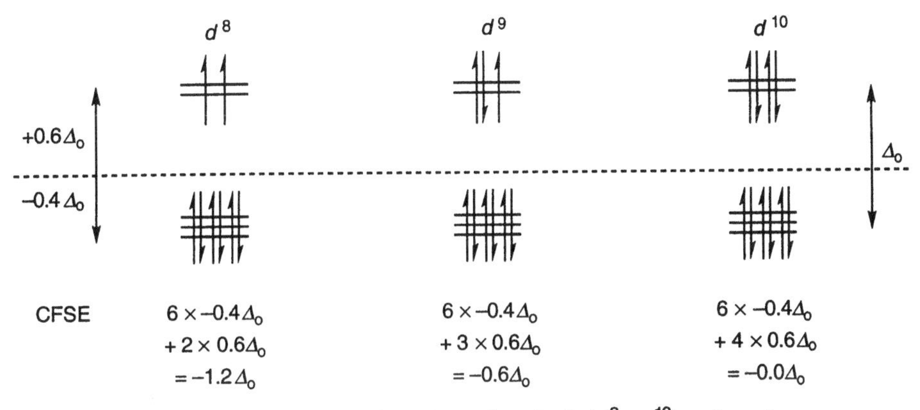

Fig. 5.16. The electronic configurations of octahedral $d^8 - d^{10}$ configurations.

The $d^8 - d^{10}$ configurations

At the d^8 configuration (Fig. 5.16), there is no such thing as a high or low spin configuration. Assuming a low spin configuration, the first six electrons would fill the t_{2g} level and the remaining two would go into the e_g level, one electron to each. Assuming a high spin configuration, the first five electrons would go one each into the five *d* orbitals and the remaining three would fill the t_{2g} set. Either way, the result is the same. Since there is only one possible configuration, the high or low spin qualifier is not to be used. Similarly, there is only one possible configuration for each of the d^9 and d^{10} configurations.

5.7 Tetrahedral complexes

The procedure for tetrahedral complexes parallels that for the octahedral case. However, recall that Δ_t is less than half the size of Δ_o. The effect of this is that the pairing energy P is *virtually always* larger than the splitting between the e and t_2 energy levels. Virtually all tetrahedral complexes are high spin. The configurations of the high spin cases and their CFSEs are shown in Figs 5.17 – 5.19.

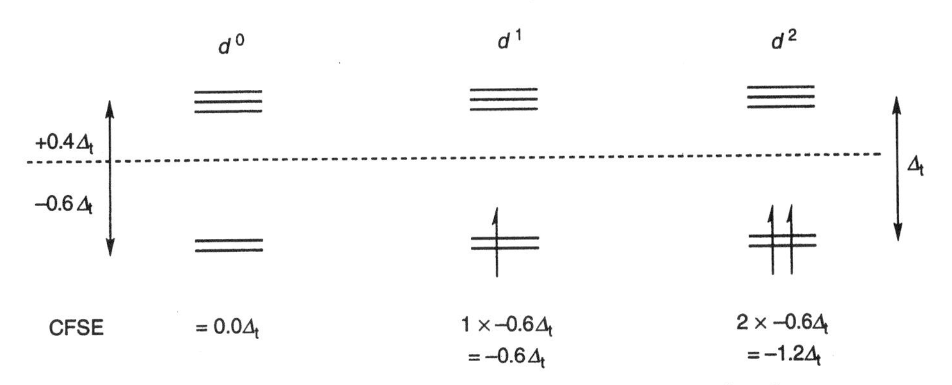

Fig. 5.17. *d* Level configurations and CFSE calculations for tetrahedral $d^0 - d^3$ complexes.

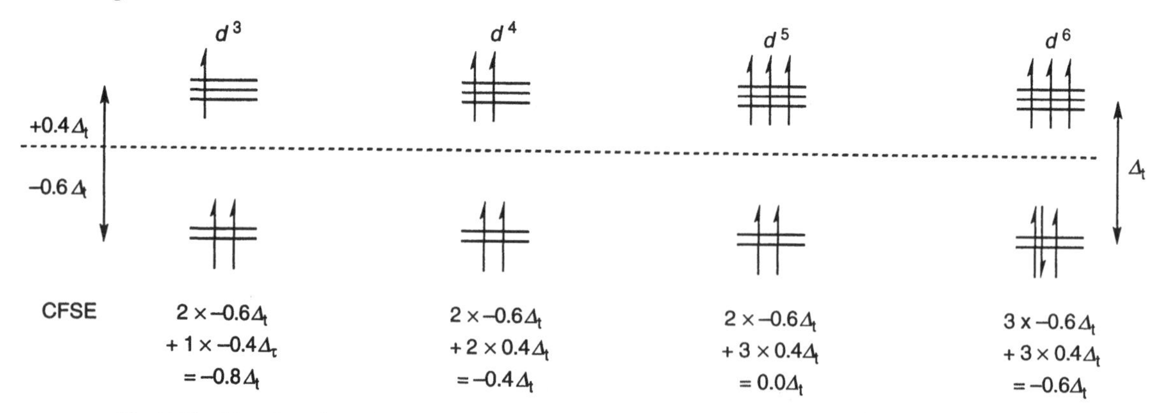

Fig. 5.18. *d* Level configurations and CFSE calculations for high spin tetrahedral $d^3 - d^6$ complexes.

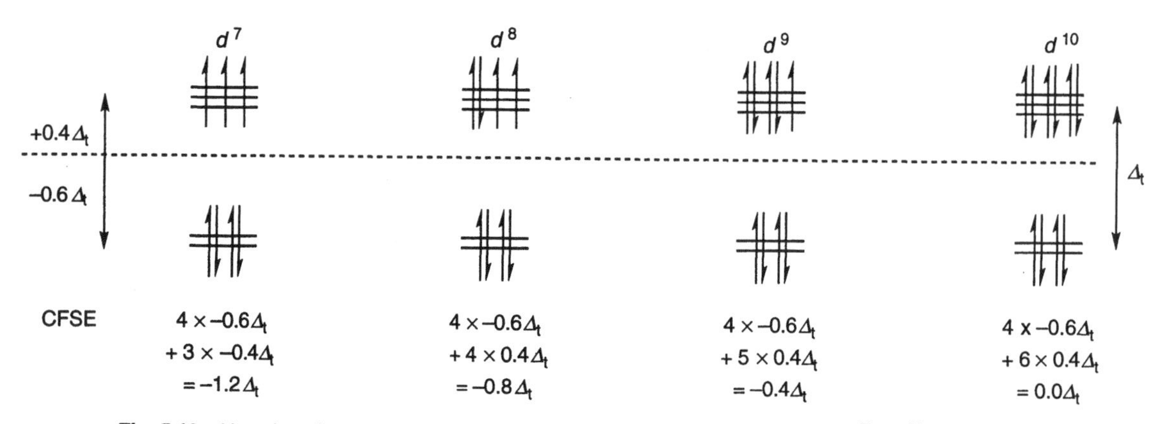

Fig. 5.19. *d* Level configurations and CFSE calculations for high spin tetrahedral $d^7 - d^{10}$ complexes.

5.8 Problems with the crystal field theory

Although the crystal field approach has its problems, this does not detract from its usefulness. Ligands are not point charges. Anions such as Cl^- are polarized away from a spherical charge distribution on being placed next to a positive charge (the metal cation). The overall charge can therefore no longer be represented as a monopole charge. An ionic approach is especially good if the respective elements are either very electropositive or very electronegative. Most transition metals and ligands are intermediate in nature. All these points require the introduction of some covalency into the model (a molecular orbital approach). A combined approach using the ionic and covalent models is referred to as the ligand field theory.

5.9 Exercises

1. Find six examples of square planar complexes. For each, determine the combined number of metal valence and donated ligand electrons. Rationalize your observations.
2. Calculate the CFSE values for *low* spin tetrahedral complexes. In which cases are the high spin and low spin configurations identical?

6 Bonding: covalent models

It was intimated earlier that one way to regard the ligand–metal interaction is as a dative covalent bond from the ligand to the metal. Ligands such as H_2O, NH_3, CO, or Cl^- are perfectly reasonable electron pair donors. There must be an orbital *into which* the electron pair donates. So which orbitals does the metal use to accept electrons? One simple model involves donation into empty metal hybrid orbitals. This is a covalent valence bond model. This model has its problems which are at least partly resolved through the use of a more sophisticated covalent model called molecular orbital theory.

6.1 A valence bond model: octahedral 6 coordination

Consider an octahedral metal complex ML_6 where L is some two electron donor ligand. Six ligands require six M acceptor orbitals to bind them. Very many transition metal complexes are octahedral. Clearly if the metal is to be hybridized to interact with six ligands, then six metal orbitals are required. The six that are used for octahedral metal hybridization (for a first row metal) are the 4*s*, the three 4*p*, and two of the 3*d* orbitals (the d_{z^2} and the $d_{x^2-y^2}$ orbitals). It makes sense to use the d_{z^2} and the $d_{x^2-y^2}$ orbitals since these are the two that point directly towards the ligands. The six hybrids are illustrated in Fig. 6.1. They are arranged in an octahedral fashion. One way to represent this hybridization is shown on the left of the energy level diagram in Fig. 6.2.

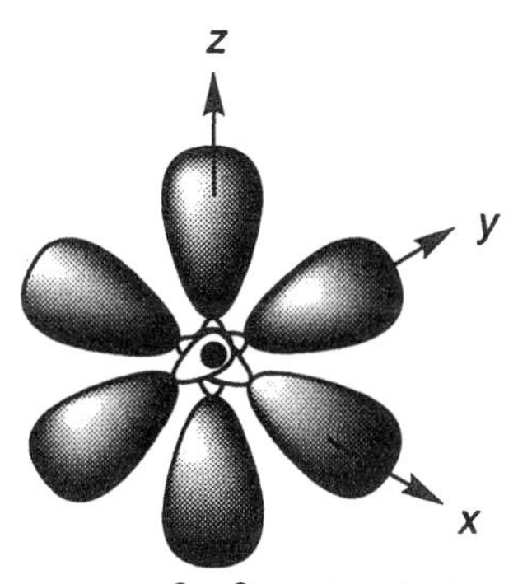

Fig. 6.1. d^2sp^3 Hybrid orbitals. They are orientated along the six axes of the octahedron, which are coincident with the *x*, *y*, and *z* axes.

The metal has nine valence orbitals. The six ligands, assuming for the present no multiple bonding (that is, the ligands are σ-bonding only with no π-effects), contribute six orbitals. There is therefore a total of fifteen valence orbitals of interest in the complex. These combine to form six bonding and six antibonding orbitals. The remaining three orbitals are nonbonding (Fig. 6.2). Mathematically, functions describing each of the hybrid and donor orbitals are mixed to form bonding and antibonding orbital functions. This accounts for the six bonding and six antibonding combinations in Fig. 6.2. Effectively, the lone pairs in the six ligand orbitals donate electrons into the six empty d^2sp^3 hybrids. In this model, the remaining three 3*d* orbitals are unaffected and so a correlation line is drawn horizontally to show there is no interaction with any other orbitals. This set of orbitals corresponds to the t_{2g} set of orbitals. Only the twelve ligand electrons are shown on this diagram. They occupy the six metal–ligand bonding orbitals. This illustrates that the nature of the bonding is dative. Each of the six bonding energy levels represents one dative bond.

The energy level diagram shows that the nature of the six bonding orbitals is dative. The twelve electrons from the ligands occupy the six lower bonding orbitals. The only other electrons of concern are the metal *d* electrons. Above the six bonding orbitals, the next set of available orbitals available for occupancy are the t_{2g} orbitals. This is where the metal *d* electrons start

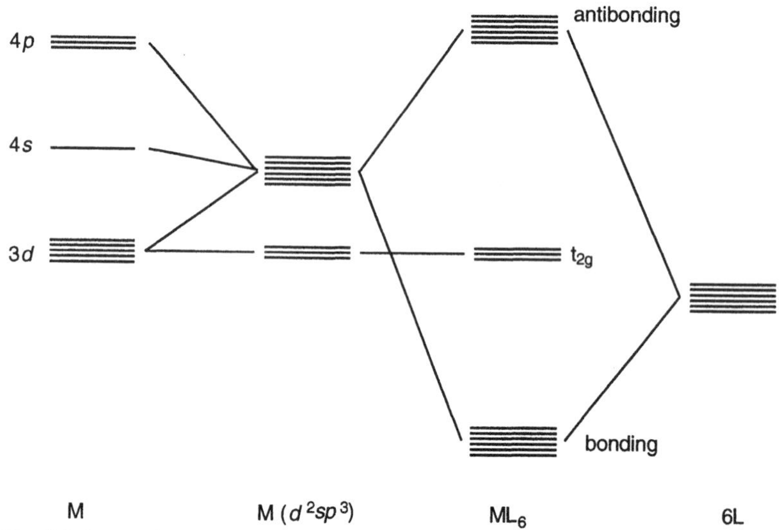

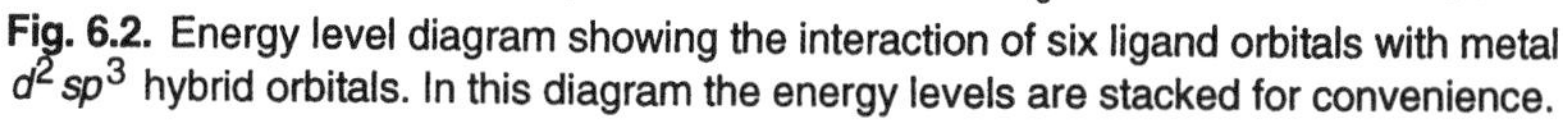

Fig. 6.2. Energy level diagram showing the interaction of six ligand orbitals with metal d^2sp^3 hybrid orbitals. In this diagram the energy levels are stacked for convenience.

feeding in. If the complex is, say d^4, then the occupancy of the t_{2g} set of orbitals is $t_{2g}{}^4$.

In practice, a *d*-block metal is much less bothered about whether its nonbonding orbitals are filled or not when compared to main group compounds, for which it is unusual for the nonbonded orbitals to be vacant. This means that this nonbonded set of orbitals can contain anything from 0 to 6 electrons. Also, electrons may move in and out of these nonbonded orbitals with some ease. In fact, this feature is in part responsible for the multiple oxidation states displayed by transition metal complexes. Reduction and oxidation corresponds to electrons being moved into and out of these approximately nonbonding orbitals, with little loss or gain of energy.

It is not unreasonable that in an octahedral complex, all the six bonding orbitals must be occupied in order to bind the six ligands. If for any reason electrons are removed from these orbitals, the result will be loss of a ligand to form a lower coordination number. Twelve valence electrons is therefore the lower allowed limit for stable electronic configurations of octahedral complexes $[ML_6]^{n+}$. The three nonbonding orbitals can hold anything from 0 to 6 electrons, meaning that valence electron counts from 12 to 18 are expected. When all three nonbonding orbitals are completely occupied, an 18-electron configuration is achieved. It will become clear later that the 18-electron configuration is highly favoured for certain classes of compound because secondary interactions cause the nonbonding set of orbitals to be lowered in energy, making it desirable that these orbitals should be filled.

This viewpoint of the bonding has its deficiencies and in practice each set of six bonding and six antibonding orbitals is not degenerate. Without some modification, this model does not account for magnetic or electronic properties of *d*-block compounds. Instead a more sophisticated covalent model is required, the *molecular orbital* method.

6.2 The molecular orbital approach for σ-bonded ligands

In the localized approach to bonding, two electrons are placed between two atoms to form a bond. In the molecular orbital approach it is recognized that there is no need to limit the orbital to just two centres. Instead the *molecular orbital* (MO) might spread over a number, or even all, of the atoms in a molecule. The reader is directed elsewhere for a full treatment of the molecular orbital approach. This section treats the bonding in a complex as though all the bonds possess σ symmetry. In Sections 6.4 and 6.5 the possibilities of π bonding are addressed.

The idea is to take each of the available metal valence orbitals (for a first-row transition metal the valence orbitals are the $4s$, $4p$, and $3d$ orbitals) and to determine what *combinations* of the ligand lone pair orbitals are allowed to overlap with the metal orbitals on grounds of symmetry. A mathematical treatment known as *group theory* is suitable for delineation of the allowed combinations. When there are just two atoms to consider, as in a diatomic molecule, it is usually intuitively clear which orbitals are allowed to overlap. Figure 6.3 shows some overlaps which are allowed on grounds of symmetry whereas Fig. 6.4 shows some that are disallowed on grounds of symmetry (that is, the value of the overlap integral = 0). In Figure 6.3 there are clear areas of overlap in the $d + s$, and $d + p$ combinations *when orientated as shown* and the overall effect is bonding. For the $p + p$, and $d + s$ combinations orientated as shown in Fig. 6.4 there is no net overlap. There are regions where there is in-phase overlap, but each is precisely countered by a corresponding area of out-of-phase overlap. The net result is zero overlap. For systems larger than two atoms the available overlaps become less clear in the absence of a group theory analysis.

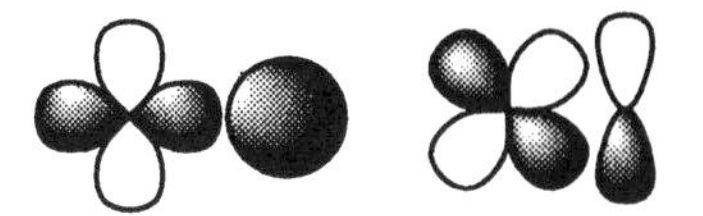

Fig. 6.3. Some orbital overlaps allowed on symmetry grounds.

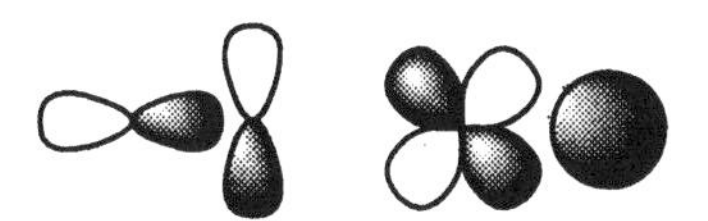

Fig. 6.4. Some orbital overlaps not allowed on symmetry grounds.

The $4s$ orbital overlaps with *all six* of the ligand lone pair orbitals to give a bonding combination (Fig. 6.5). This is the a_{1g} combination, (a group theory label). There is a matching anti-bonding combination, not shown here, also labelled a_{1g}. Mathematically the antibonding combination is derived by reversing the signs of the ligand orbitals when formng these combinations.

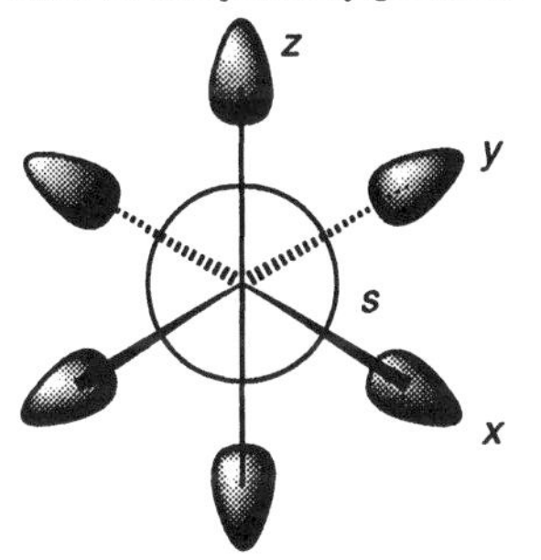

Fig. 6.5. The a_{1g} bonding combination from overlap of a metal s orbital with six ligand σ-orbitals in an octahedral array.

The situation for the p orbitals is shown in Fig. 6.6. Each of these combinations is also bonding. This group of three orbitals taken together is called the t_{1u} set of orbitals. They are grouped together in this fashion because each of these combinations is degenerate (symmetrically equivalent to the other two).

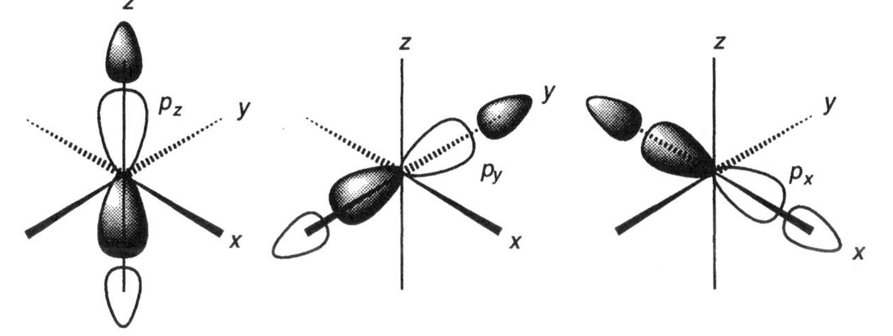

Fig. 6.6. The overlap of metal p orbitals with ligand combinations of donor orbitals.

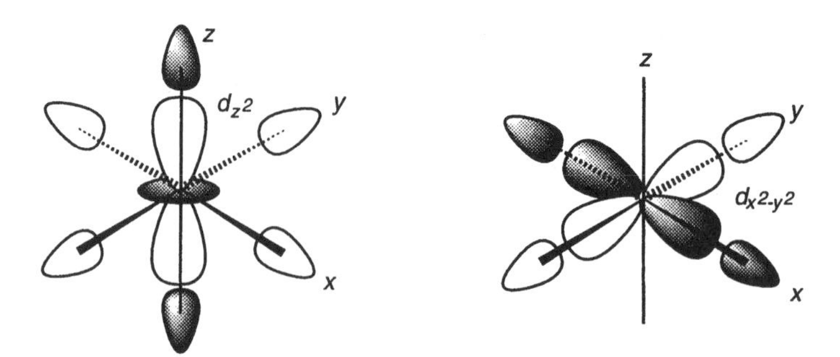

Fig. 6.7. The overlap of metal d_{z^2} and $d_{x^2-y^2}$ orbitals with combinations of ligand donor orbitals.

It is best to consider the *d* orbitals in two groups. The first group contains the two orbitals pointing directly at the six ligand lone pairs. These are the $d_{x^2-y^2}$ and d_{z^2} orbitals. They form bonding combinations with the lone pairs. The bonding combinations are shown in Fig. 6.7. Taken together, they are called the e_g set of orbitals. Their energies are degenerate, even though the pictorial representations look very different. Again this is related to the mathematical function describing the d_{z^2} orbital, which is a linear combination of the mathematical functions $d_{z^2-x^2}$ and $d_{z^2-y^2}$.

This leaves the three remaining *d* orbitals. These are the three pointing between the ligands. It is not possible to write down any bonding combination for these three orbitals. Try it! The problem is illustrated in Fig. 6.8. No matter what sign is given to any combination of the ligand orbitals, the net result is always zero overlap. This means that, by symmetry, the d_{xy}, d_{xz}, and d_{yz} orbitals are all nonbonding.

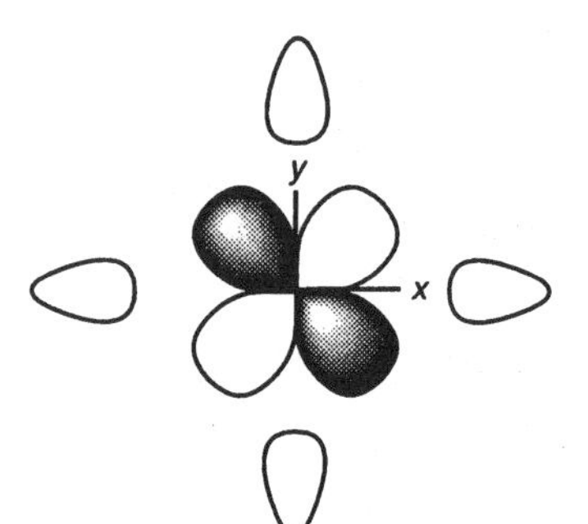

Fig. 6.8. The relative positions of the metal d_{xy} orbital and four ligand *s* orbitals.

There are therefore six ligand σ orbitals involved in bonding with just six of the metal valence orbitals. This leads to the formation of six bonding orbitals and six antibonding orbitals. An energy level representation for this situation is shown in Fig. 6.9. This diagram should be compared with Fig. 6.2 derived by the valence bond approach. In both cases there are six bonding and six antibonding levels, but in the MO energy level diagram the six bonding orbitals do not form a degenerate set. Instead there are three levels, one of which is triply degenerate (t_{1u}), one of which is doubly degenerate (e_g), and one of which is singly degenerate (a_{1g}). There is also a relationship between the crystal field splitting diagrams derived earlier for an octahedral field and Fig. 6.9. The nonbonding t_{2g} and the weakly antibonding e_g sets of orbitals in Fig 6.9 correspond to the t_{2g} and the e_g orbitals shown in the crystal field energy splitting diagrams.

6.3 Other geometries

The previous section, in essence, quotes graphically the results of a detailed group theory analysis for the octahedral ML_6 system. Related analyses are possible for other geometries such as the tetrahedron, trigonal bipyramid, etc. Note that in the analysis for the ML_6 system the ligand σ orbitals are unable on symmetry grounds to interact with any of the three t_{2g} orbitals. These

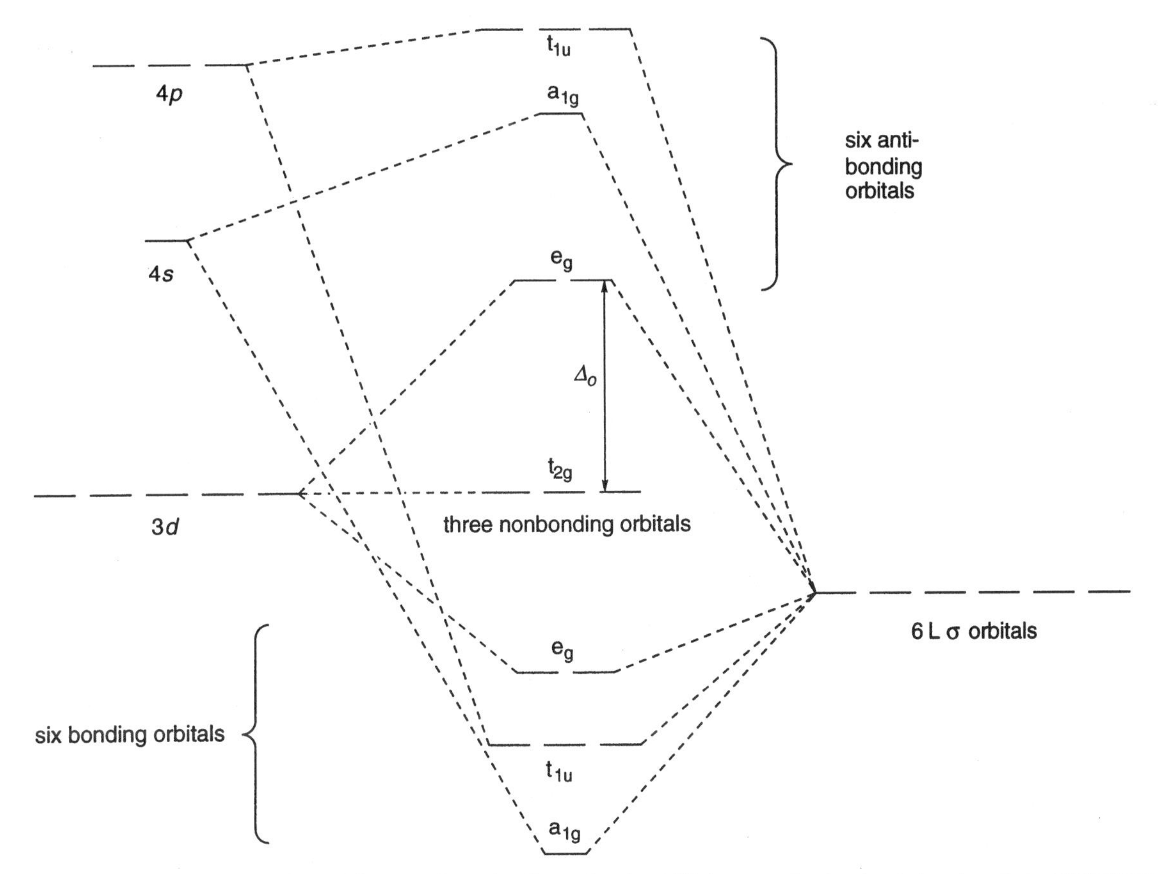

Fig. 6.9. Energy level diagram for an ML_6 complex using the molecular orbital approach, showing only σ-bonding contributions.

orbitals (d_{xy}, d_{xz}, and d_{yz}) are therefore *strictly nonbonding* (in the absence of π interactions, discussed in the following section).

A related phenomenon shows up in an analysis of the cubic ML_8 system. In this case it is the d_{z^2} and the $d_{x^2-y^2}$ orbitals which are unable to interact on symmetry grounds with any combination of the metal σ orbitals. Combinations of all the other seven orbitals (one *s*, three *p*, and the remaining three *d* orbitals) are allowed by symmetry arguments to interact. However, this means that there are only seven metal orbitals available for bonding to eight ligands, which while not disqualifying the *cubic* ML_8 geometry for *d*-block elements does tilt the balance in favour of more commonly observed geometries. This restriction is not present for the *f*-block, some of whose complexes do display a cubic geometry, since there are *f* orbitals which *are* permitted on symmetry grounds to interact with the eight ligand σ orbitals.

6.4 π-Donor ligands

Thus far, the bonding of a ligand to a metal has been treated in both the ionic and covalent models as a pure σ-donor interaction involving just an appropriate ligand σ-donor orbital. However, in many cases the bonding of a ligand to a metal also involves π-interactions, that is, there is an element of

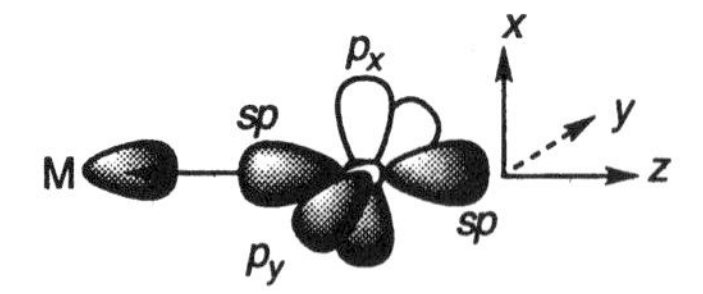

Fig. 6.10. Hybridization of halide in a metal–halide complex. All halide orbitals are filled.

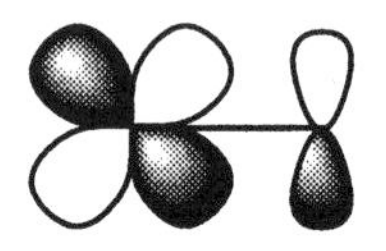

Fig. 6.11. The interaction of one of the metal t_{2g} orbitals with a single ligand *p* orbital.

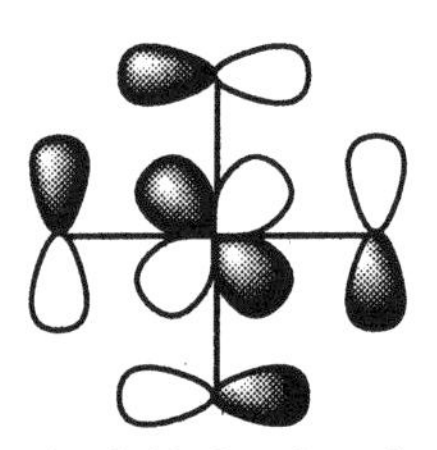

Fig. 6.12. Overlap of a metal t_{2g} *d* orbital with a linear combination of four halide *p* orbitals to make a five centred molecular orbital.

multiple bonding between the ligand and the metal. The π effects may be subdivided into π-donor effects and π-acceptor effects. This section addresses the cases where in addition to these σ-donor interactions, π-donor bonding also plays a part. The most common examples of π-donor interactions involve halide ligands. This discussion addresses octahedral complexes but the same principles apply for other geometries.

It is convenient to regard a halide ligand X^- as *sp* hybridized (Fig. 6.10). One of the *sp* hybrids is the lone pair used in the M—X σ-donor bond. The other hybrid is located on the other side of the M—X interaction and effectively plays no further part in bonding. The remaining lone pairs are halide atomic p_x and p_y orbitals. All the halide orbitals are filled. The filled halide p_x and p_y orbitals are of the correct symmetry to interact with some of the metal *d* orbitals (Fig. 6.11).

Strictly, when analysing the bonding in a metal complex it is necessary to treat the molecule as a whole. In an octahedral complex ML_6, it is the t_{2g} *d* orbitals which are able to interact with the ligand *p* orbitals. Each of the six halide ligands has one p_x and one p_y orbital available for π-overlap with the metal *d* orbitals, making twelve in all. Figure 6.12 shows the interaction of one of the metal t_{2g} *d* orbitals with a linear combination of four of these twelve orbitals (also labelled t_{2g}). The resulting interaction is a five-centred molecular orbital. There are three equivalent such combinations in each of the *xy*, *xz*, and *yz* planes. The filled halide *p* orbitals are *lower* in energy than the metal t_{2g} orbitals and are occupied. The energy level diagram to describe the interaction between the two sets of orbitals is shown in Fig. 6.13.

This shows that the interaction of the two sets of t_{2g} orbitals leads to a bonding and an antibonding combination. The lower of these is closer in terms of energy to the ligand *p* orbitals and is therefore largely ligand based. The upper set (the antibonding combination) is closer in energy terms to the metal t_{2g} *d* orbitals and so is largely metal based. Effectively, the metal t_{2g} orbitals are pushed *up* in energy a little. These orbitals may or may not be occupied but any such occupation is not shown in the Figure. This set of orbitals is best regarded as a modified set of metal orbitals and so the new value of Δ_o is the difference in energy between this set and the untouched metal e_g set of orbitals. By definition, this causes Δ_o to *decrease*. The

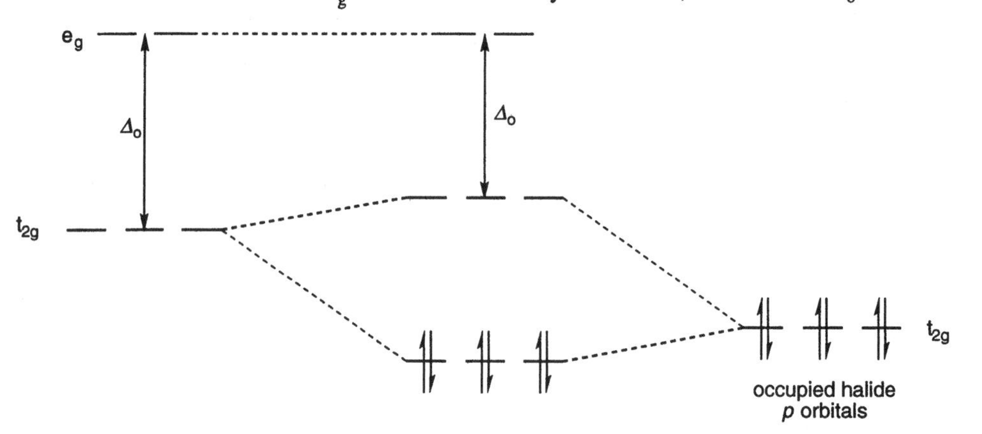

Fig. 6.13. The interaction between empty metal t_{2g} orbitals and filled halide *p* orbitals in an octahedral environment.

interaction of the two sets of orbitals corresponds to a partial transfer of electron density from ligand *p* orbitals to the metal, hence the term π donor.

6.5 π-Acceptor ligands

Halides are good examples of π-donor ligands. They possess *filled* orbitals with the correct symmetry to interact with some of the metal *d* orbitals. There are a number of other ligands which have *vacant* orbitals with the correct symmetry to interact with some of the metal *d* orbitals.

Carbon monoxide is an excellent example of a π-acceptor ligand. In this case it is the carbonyl π^* orbitals which are of the correct symmetry to interact. As for the halides discussed in the previous section, there are two types of interaction to consider. The first is a σ-donor interaction from the CO carbon lone pair into an empty metal orbital. Looked at simplistically the triple bond of CO is made up from a σ bond and two π bonds. The π orbitals of CO are filled but it is the *empty* π^* orbitals which are of interest here. For convenience the situation is discussed here just for an octahedral complex. The arguments are, however, similar for other geometries. The metal t_{2g} orbitals (the d_{xy}, d_{xz}, and d_{yz} orbitals) of the octahedral metal complex are normally filled in this type of complex. The π^* orbitals are of the correct symmetry to overlap with the metal t_{2g} orbitals. This is shown for a single CO ligand in Fig. 6.14. There is an exactly equivalent interaction involving the second π^* orbital at right angles to that shown. The six CO ligands between them possess twelve π^* orbitals. Again, strictly, the complex should be considered as a whole and the reader is left to construct a diagram for carbonyl bonding analogous to Fig 6.13 to reflect this.

In this situation, the metal t_{2g} orbitals would normally be filled. The interaction between the metal t_{2g} orbitals and linear combinations of the CO π^* orbitals effectively means that there is a flow of electron density from the metal to the ligand (Fig. 6.15). The donation of electrons from the metal to the ligand is referred to as '*back-bonding*' since the direction of electron transfer (M$\rightarrow$L) is the opposite to that normally seen (L$\rightarrow$M). The symmetry of the overlap is clearly π-type, hence the term π back-bonding. Recall the definitions of the terms *Lewis acid* and *Lewis base* which involve the terms '*electron acceptor*' and '*electron donor*' respectively. Therefore the ligand CO is quite reasonably referred to as a 'π-acid' as well as a 'σ-base'. To correspond with this, halides are π-bases as well as σ-bases.

Metal	π Lewis acid or base
	σ Lewis acid
CO	π Lewis acid
	σ Lewis base
Halide	π Lewis base
	σ Lewis base

To sum up, *carbon monoxide acts as a σ-donor and a π-acceptor.* When

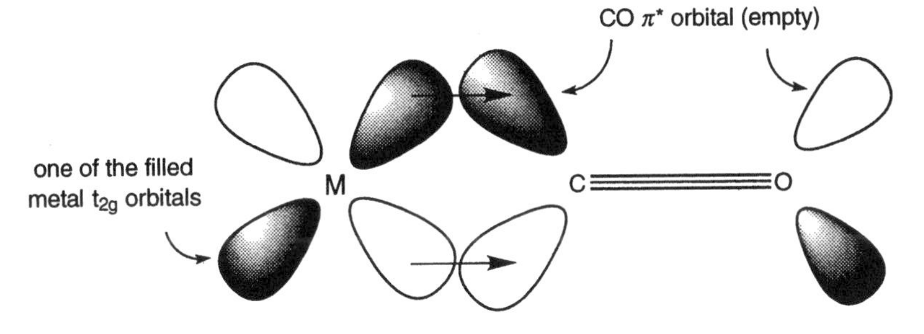

Fig. 6.14. One of the π-bonding contributions in a M—CO bond.

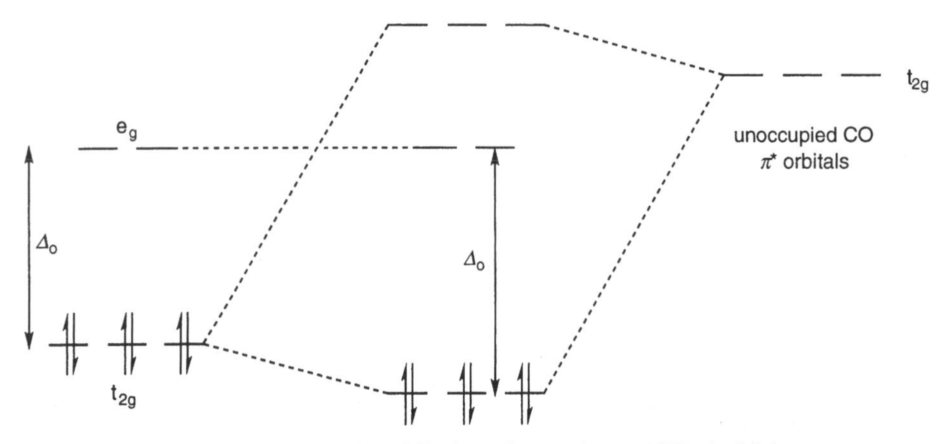

Fig. 6.15. The interaction of filled metal t_{2g} and vacant CO π^* orbitals.

the bonding situation is such that electrons are transferred in one direction and other electrons accepted in the reverse direction, the bonding is said to be synergistic, (*synergy,* Gr. synergia, *n.* combined or coordinated action). The bonding is cooperative in the sense that electrons transferred in one direction would lead to an unacceptable charge build up at the other centre, but for the removal of some electron density in the reverse direction, and so on. Up to a point the σ-interaction strengthens the π-interaction, and *vice versa.*

Overall, there is only a small dipole moment associated with the M—C interaction, suggesting approximate electroneutrality. However, the effect of the electronegative oxygen atom is to cause a small charge imbalance between carbon, which becomes slightly positively charged, (δ^+) and oxygen which becomes slightly negatively charged (δ^-). This renders the carbon atom somewhat prone to attack by nucleophiles (such as hydride donors) and the oxygen by electrophiles (such as $AlCl_3$).

The effect of the orbital interactions illustrated in Fig. 6.15 upon the metal t_{2g} orbitals is to produce a set of bonding orbitals with the predominant appearance of metal t_{2g} orbitals but *lower* in energy. In the d^6 situation, these orbitals are filled and since the energy of these electrons drops, the overall effect is to increase the total bonding of the system. Note that in the σ-bonding only case (Fig. 6.9), the metal t_{2g} orbitals are nonbonding but for a complex with π-acceptor ligands are bonding. The π-interaction has no effect upon the e_g orbitals since there are no ligand orbital combinations with e_g symmetry. The net effect of a π-acceptor ligand is therefore to increase the splitting between the e_g and t_{2g} orbitals, that is, to *increase* Δ_o.

Evidence for π-acceptor properties of CO

Multiple bonds between elements tend to shorten the internuclear distance. Thus a C—C bond is about 154 pm in length whereas a C=C bond is 134 pm. The back bonding phenomenon for metal carbonyl complexes corresponds to an increase in bond order between the metal and the carbonyl carbon atom. Therefore the metal–carbon bond should be shorter than a M—C single bond. The X-ray crystal structures of metal carbonyl complexes such as $[Cr(CO)_3(dien)]$ (Fig. 6.16) provides evidence for this bond shortening. Each of the nitrogen atoms in this complex is sp^3 hybridized and sp^3 hybridized nitrogen has an atomic radius of 70 pm. Each carbonyl carbon is sp hybridized and has exactly the same atomic radius as sp^3 nitrogen. However although all the ligand atoms are the same size, the chromium–carbon bond lengths are all somewhat shorter than the chromium–nitrogen bond lengths. Assuming the chromium is spherical, this is good evidence for multiple bond character in metal–carbonyl interactions. Similar arguments based upon M—L bond lengths apply to other π-acceptor ligands (next Section).

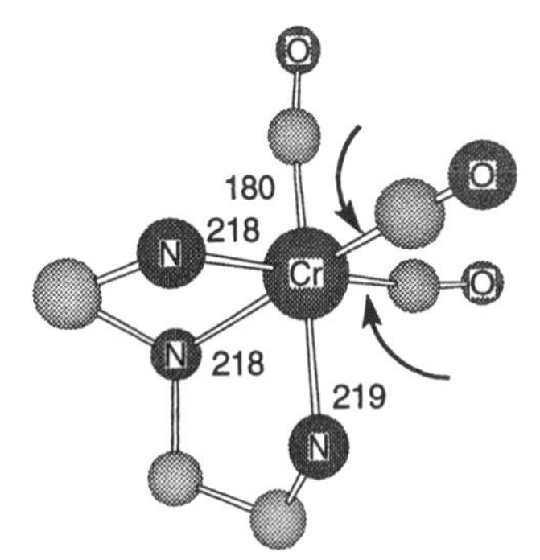

Fig. 6.16. The solid state structure of [$Cr(CO)_3$(dien)]. Bond lengths expressed in pm.

6.6 Other π-acceptor ligands

Carbon monoxide is a molecule containing carbon (s^2p^2, four valence electrons) and oxygen (s^2p^4, six valence electrons): a diatomic species with a total of ten valence electrons. Other diatomic molecules with ten valence electrons are said to be isoelectronic with CO. Apart from the relatively rare CS, CSe, and CTe, there are several other possibilities (Table 6.1) and each will be discussed briefly.

Table 6.1. Ligands isoelectronic with CO.

ligand	formal constitution	valence electron total
CN^-	$C(s^2p^2) + N^-(s^2p^4)$	10
N_2	$N(s^2p^3) + N(s^2p^3)$	10
NO^+	$N(s^2p^3) + O^+(s^2p^3)$	10

Cyanide

Cyanide, CN^-, is a very good σ-donor but it is not necessary to invoke back bonding when it is complexed to metals in oxidation states such as II or III. Typical complexes of this type include $[Fe(CN)_6]^{4-}$ and $[Ni(CN)_4]^{2-}$.

Because of its electronic similarity to CO, cyanide is also capable of functioning as a π-acceptor ligand under some circumstances. In fact, it is not as good a π-acceptor as CO. In part, this is attributable to the negative charge. In addition, the π^* orbitals of cyanide are rather higher in energy than the π^* orbitals of CO. This means that CN^- is a poorer π-acceptor than CO because raising the energy of the π^* orbitals renders the overlap between them and the appropriate nonbonding metal *d* orbitals poorer, and therefore weaker.

Dinitrogen complexes

Dinitrogen, N_2, is also isoelectronic with CO. As such, one might expect there to be many N_2 complexes in the literature. In practice, there are relatively few. The relative energies of its orbitals renders N_2 a weaker σ-donor than CO, as well as a weaker π-acceptor. The first N_2 complex was synthesized in 1965 from the reaction of aqueous $RuCl_3$ with hydrazine (Fig. 6.17). Sometimes, N_2 can act as a bridging ligand between two metals. Other examples of dinitrogen complexes are shown in Fig. 6.18.

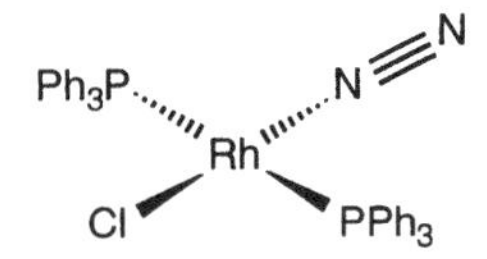

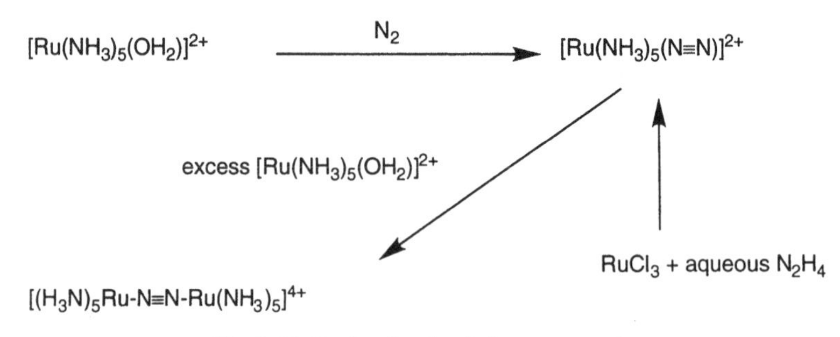

Fig. 6.17. Synthesis of a dinitrogen complex.

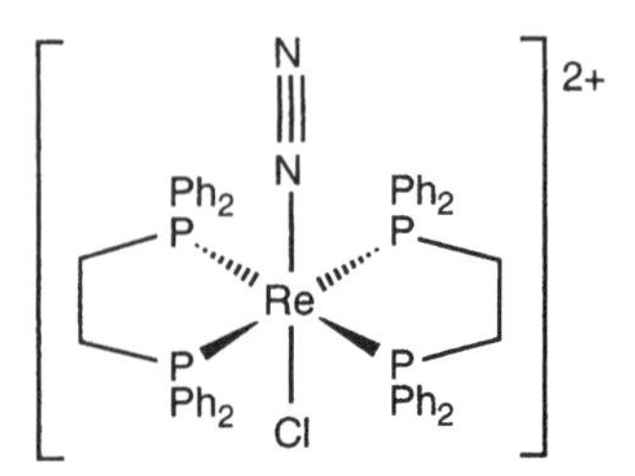

Fig. 6.18. Other examples of metal dinitrogen complexes.

Dinitrogen complexes have a very real importance. It has been recognized for some time that a class of metalloenzymes known as nitrogenases catalyse the reduction of N_2 to ammonia under ambient conditions in aqueous media. This would be a very valuable process if one could accomplish this on an industrial scale under mild conditions. The biological process involves bacteria in the root nodules of plants such as clover and is iron and molybdenum based, (molybdenum is the only second row *d*-block element demonstrated to have a rôle as a necessary trace element).

A few examples are known in which dinitrogen is converted into ammonia from relatively simple dinitrogen complexes (Fig. 6.19). The interesting thing about this dinitrogen complex is that treatment with acid causes evolution of ammonia, effectively a nitrogen fixation. Experiments such as this are interesting as they might be 'model systems' for the biological fixation of nitrogen.

THF

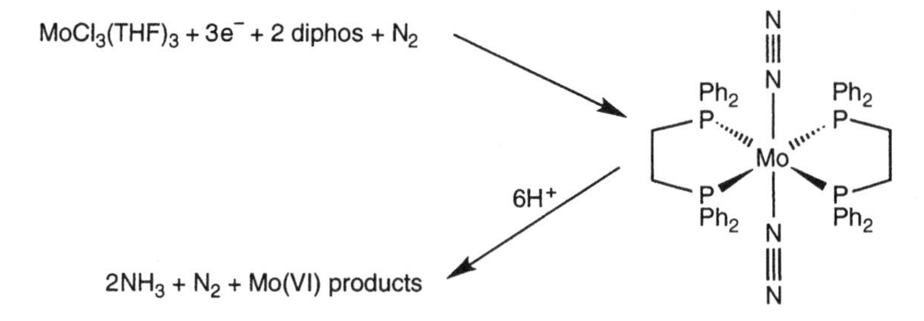

Fig. 6.19. Formation of NH_3 in the laboratory.

6.7 Nitrosyl complexes

Both $[Cr(NO)_4]$ and $[Mn(CO)(NO)_3]$ are isolelectronic and isostructural with $[Ni(CO)_4]$

The nitrosonium ion, NO^+, is isoelectronic with CO. It is therefore perhaps not surprising that there are many complexes of NO^+ (nitrosyl) that are both isoelectronic and isostructural with carbonyl complexes. For the purpose of the oxidation state formalism, the ligand is regarded as NO^+ ($:N{\equiv}O^+$) since this leaves both atoms with an octet of electrons. This leads to very low *formal* oxidation states (some examples of oxidation state determinations are shown in Table 6.2) but recall again that oxidation state is a formalism and does not have any relation to the actual charge on the metal. In metal nitrosyl complexes the M–N distance is *shorter* than the sum of the covalent radii of the metal and nitrogen, indicating some back-bonding.

The importance of nitrosyls is partly to do with atmospheric nitrogen oxides. They are harmful and enter the environment through, for instance, car exhausts. This will no doubt lead to further interest in their chemistry. Ruthenium in particular seems to have a marked affinity for nitrosyls and this is one of the reasons why it is incorporated into catalytic converters.

There is a complication for nitrosyl complexes. In some complexes (Fig. 6.20), the nitrosyl group bonds in a quite different fashion. The alternative geometry has a *bent* M—N—O interaction. Typical values for the M—N—O bond angle in such cases are 120–140°. This is a *distinct structural difference* from the more usual linear nitrosyls and does not arise from solid state packing effects. For oxidation state formalism purposes a bent nitrosyl is regarded as a complex of NO^- ($:N^-=O$, again both atoms are in octet configurations).

Figure 6.21 shows the formal derivation of linear and bent nitrosyls from neutral NO and indicates the oxidation state conventions for NO ligands. Formally, the transformation of a linear nitrosyl into a bent nitrosyl is an internal transfer of a pair of electrons from the metal, (oxidation by 2 units) to the NO^+ ligand, which is reduced by two units, making it NO^-. The nitrogen rehybridizes to sp^2, making the M—N—O bond bent.

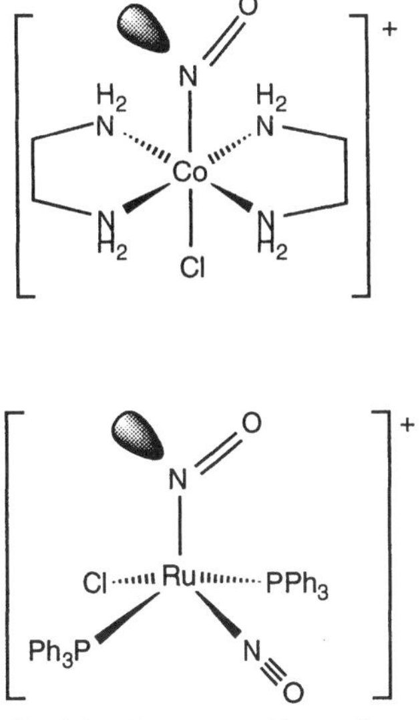

Fig. 6.20. Examples of bent nitrosyl complexes.

Table 6.2. Examples of oxidation state determinations for nitrosyl complexes.

$[Cr(NO)_4]$	**contribution**	**running total**
number of Cr(O) valence electrons	6	**6**
charge on complex: 0, assume metal centred,	0	6
remove four NO^+ ligands, therefore add 4	4	10
difference between 6 and 10 gives oxidation state		– 4

$[Mn(CO)(NO)_3]$	**contribution**	**running total**
number of Mn(O) valence electrons	7	**7**
charge on complex: 0, assume metal centred,	0	7
remove one neutral CO ligands	0	7
remove three NO^+ ligand, therefore add 3	3	10
difference between 7 and 10 gives oxidation state		**– 3**

$[RuCl(PPh_3)_2(NO)_2]^+$ (Fig. 6.20)	**contribution**	**running total**
number of Ru(O) valence electrons	8	**8**
charge on complex: +1, assume metal centred,	– 1	7
remove one anionic Cl^- ligand, therefore subtract 1	– 1	6
remove two neutral PPh_3 ligands	0	6
remove one NO^+ ligand, therefore add 1	1	7
remove one NO^- ligand, therefore subtract 1	– 1	6
difference between 8 and 6 gives oxidation state		**2**

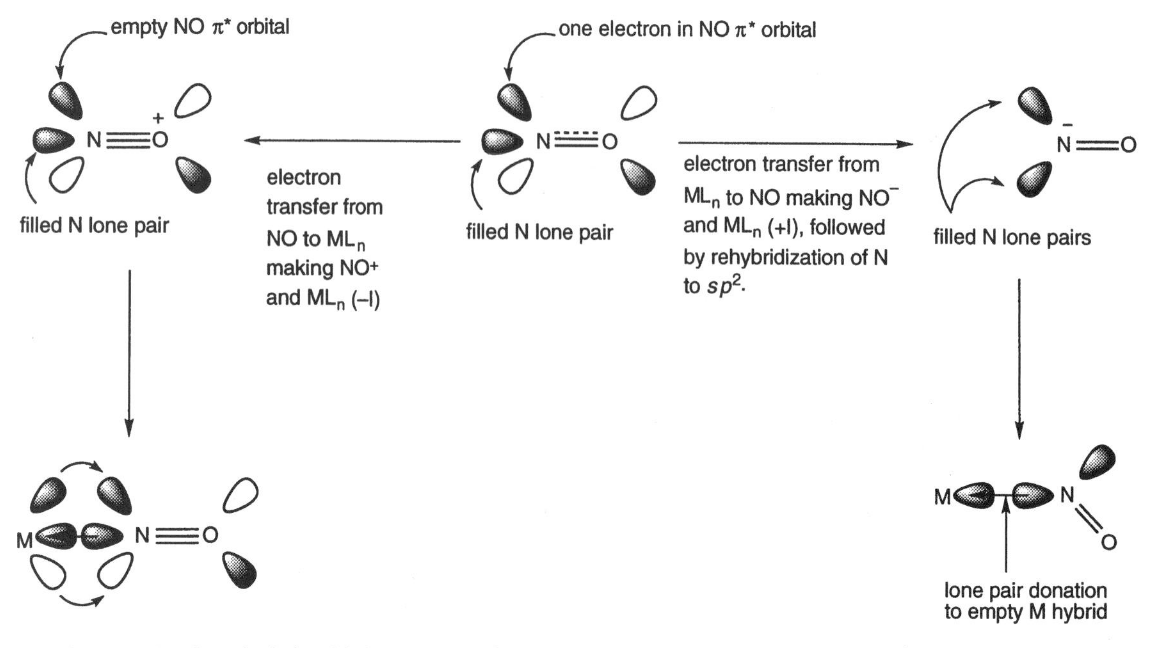

Fig. 6.21. The formal relationship between bent (complex of NO^-) and linear (complex of NO^+) nitrosyl complexes.

6.8 Phosphine and phosphite complexes

Phosphines, PR_3, and phosphites, $P(OR)_3$, are some of the most common ligands associated with transition metal complexes. One important feature is that they are *one of the few ligand types whose properties can be varied systematically by changing the R groups* (Fig. 6.22). This allows one to tune *both* the steric and electronic properties of the ligand. Phosphines have a very real industrial importance since they are frequently added to catalytic reactors to promote catalysis. The idea is that they bind to the metal centre and so exert an influence on the catalytic cycle. They do tend to undergo degradation under the reaction conditions of a reactor which does limit their usefulness to some extent.

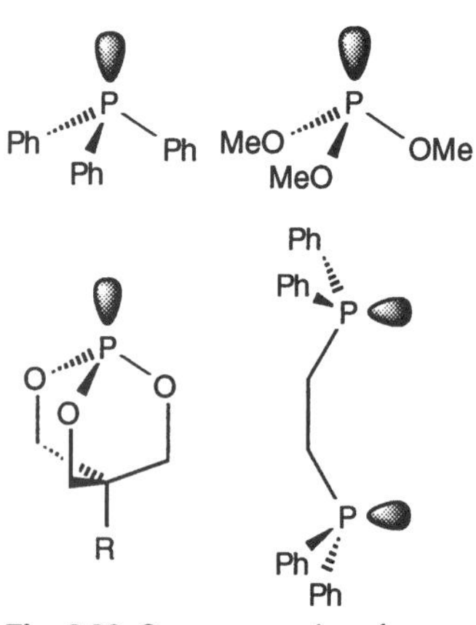

Fig. 6.22. Some examples of phosphine and phosphite ligand structural types.

Amines do not have the ability to stabilize low oxidation states through back-bonding simply because there are no empty *p* or *d* type orbitals of sufficiently low energy to accept electron density. The situation is different for phosphines. There are alternative views regarding the nature of the back-bonding interaction in metal phosphines. The classical view is that phosphorus has vacant *d* orbitals which are available for overlap with suitable filled metal orbitals (Fig. 6.23). This is in addition to donation of an electron pair from phosphorus to the metal in a σ-bond. A more recent model invokes back-bonding from the metal *d* orbitals into the P—R σ^* orbitals (the normally empty antibonding orbitals associated with the P—R σ-bonds). This effect is shown for one of the σ^* bonds in Fig. 6.23.

As the electronegativity of the R groups *increases* the π-acceptor properties of the phosphine *increase* and so other π-acceptor ligands are not required to accept as much electron density. This is illustrated by bond length

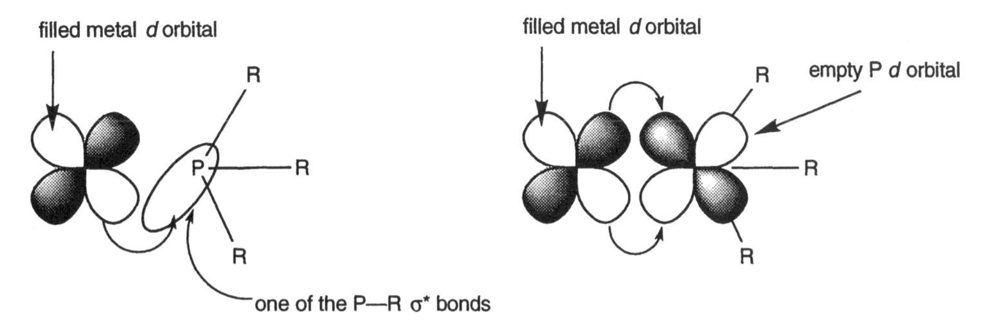

Fig. 6.23. Two views of back-bonding in metal phosphine complexes.

data for the carbonyl complexes $[Cr(CO)_5(PPh_3)]$ and $[Cr(CO)_5\{P(OPh)_3\}]$ (Fig. 6.24). The Cr—P bond shortens on increasing the electronegativity of the R substituent. This suggests that there is at least some back-bonding for the phosphite complex. Further, the *trans* M—CO bond length increases, a consequence of the way that the phosphite competes successfully for the back-bonding electron density from the metal.

It is therefore clear that electronic factors are important for phosphine complexes. However, because of the diversity of phosphine substituents, steric factors are also extremely important. For instance, the palladium phosphine complexes $[Pd(PR_3)_4]$ and $[Pd(PR_3)_3]$ are in equilibrium (Eqn 6.1).

$$[Pd(PR_3)_4] \rightleftharpoons [Pd(PR_3)_3] + PR_3 \qquad (6.1)$$

The position of this equilibrium depends very much upon the bulk of the ligand. The extent of dissociation decreases in the order: $PPhBu^t_2$ > $P(cyclohexyl)_3$ > PPr^i_3 > PPh_3 = PEt_3 > $PMePh_2$ > PMe_2Ph > PMe_3. Such equilibria are important in catalytic processes since the extent of dissociation of a phosphine ligand might determine the concentration of sites available for substrate coordination. On the other hand, although very bulky groups are effective at promoting the formation of active sites, the remaining groups may be too large to allow substrates to approach. It would be useful to have a measure of the size of a phosphine ligand. This is provided by the 'cone angle' concept.

The cone angle is one way of quantifying the size of a ligand. In practice the number is used mostly for phosphines but cone angles are also assigned to other ligands. The definition of the phosphine cone angle, θ, is given in Fig. 6.25 and related diagrams may be used to define cone angles for other ligands. The M—P distance is set at 228 pm which is generally close to the

Fig. 6.24. Bond length data for $[Cr(CO)_5(PPh_3)]$ and $[Cr(CO)_5\{P(OPh)_3\}]$.

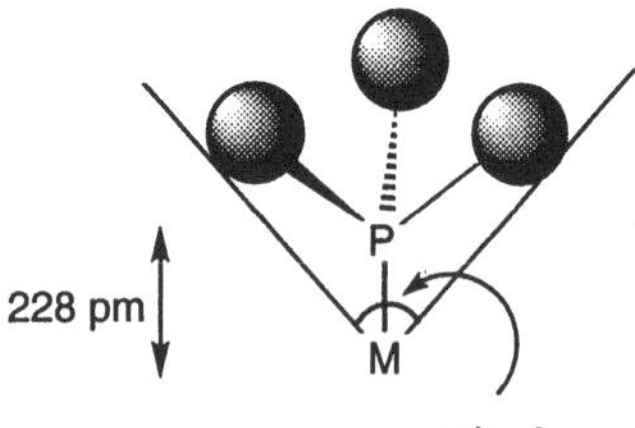

Fig. 6.25. Definition of cone angle for a phosphine. The three balls denote three arbitrary groups.

Table. 6.3. Cone angle values for some ligands.

ligand	cone angle (°)	ligand	cone angle (°)
H	75	Cl^-	102
CO	95	Br^-	105
Me^-	90	I^-	107
NH_3	94	NEt_3	150
NPh_3	166	PH_3	87
PF_3	104	PPh_3	145
$P(OMe)_3$	107	PMe_3	118
PEt_3	132	PBu^t_3	182

actual situation. The cone angle is a solid angle. Large phosphines have large cone angles.

6.9 Alkene complexes

During the first half of the 19th century (1827), Zeise prepared a platinum salt formulated as $KCl.PtCl_2.EtOH$. Its true nature was not actually recognized until the 1950s when its structure was established as an alkene complex $[K][PtCl_3(H_2C{=}CH_2)].H_2O$ (Figs 2.6 and 6.26). The alkene ligand represents another extremely important class of ligand. All the ligands considered to date in any detail bind to the metal through a single donor atom using a σ-bond. Alkenes are different, because *both* of the carbon atoms of the alkene are attached to the metal. Nevertheless, the bonding scheme is still related to that of other π-acceptor ligands in the sense that there is a σ-bonding dative component from the ligand and a π-acceptor component using π-acceptor orbitals on the alkene (Fig. 6.27).

Here the back-bonding is taking place in a *sideways* sense, involving two atoms rather than one. This is an example of *sideways* π-acid behaviour. One important consequence of this bonding scheme is that the inclusion of electrons in the π^* orbitals results in a *decrease* in the C—C bond order for the alkene. This causes the C—C alkene bond distance to *increase*, but not by very much in this particular case (from 134 pm in C_2H_4 to 137 pm in the complex). The extent to which the bond distance increases is very dependent on the particular complex. The amine complex *trans*-$[PtCl_2(NMe_2H)(C_2H_4)]$ is geometrically very similar to Zeise's salt, but has a C—C distance of 147 pm. This is getting rather close to the C—C single bond distance usually quoted as about 154 pm.

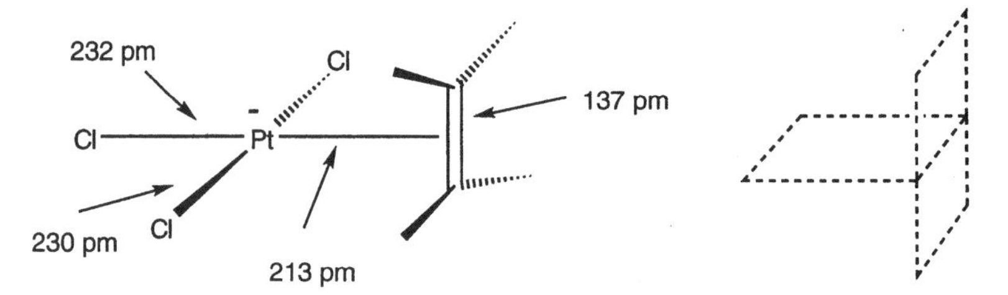

Fig. 6.26. A schematic representation of the structure of Zeise's salt.

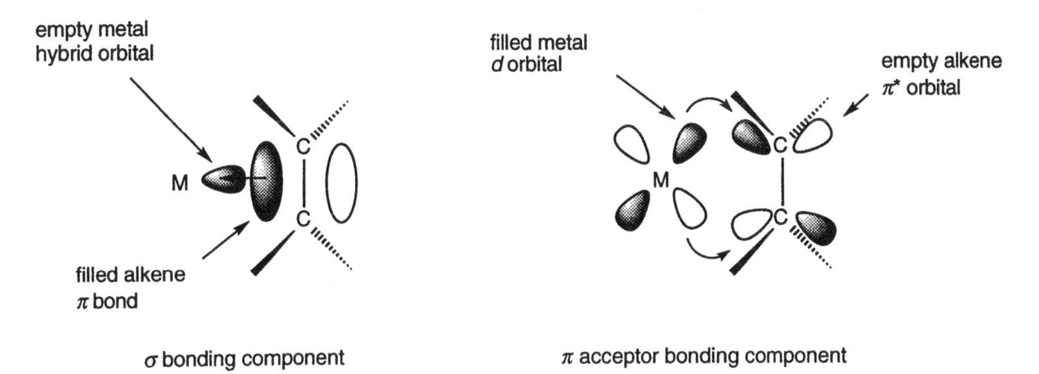

Fig. 6.27. The bonding of an alkene to a metal.

Thus, substituents on the alkene, or the metal, its oxidation state, and the other ligands all help determine the extent of back-bonding and hence C—C lengthening. As an example, electron withdrawing substituents on the alkene would be expected to promote back-bonding, and make the C—C bond length that much greater.

6.10 The eighteen-electron rule

When first row main group elements form compounds they generally achieve the electronic structure of the next noble gas, that is, neon. The reason for this is that special stability is gained if all the valence orbitals are occupied. For a first row element this corresponds to filling the single 2*s* and the three 2p orbitals. Since each orbital can hold two electrons, eight are required in total. Hence the octet rule. The eight electrons come from the element in question together with contributions from bonded atoms or groups. Generally, each covalent bond holds a single electron from each bonding partner, but there are instances in which one element donates both electrons into the covalent bond, in which case the bond is referred to as a dative covalent bond.

In essence, the rules are the same for *d*-block metal complexes. For a main group compound the four valence orbitals lead to an eight electron rule. A *d*-block metal has nine orbitals regarded as valence orbitals. For a first row *d*-block element these are the five 3*d*, the single 4s, and the three 4p orbitals. This suggests there might be a particular stability associated with the number 18. This is indeed the case. There are many examples of *d*-block metal complexes for which the 18-electron configuration is particularly stable. Complexes for which there is an overall count of 18 valence electrons associated with the metal are said to obey the 18-electron rule.

Alternative names for the 18 electron rule are the 'noble gas rule' and the 'effective atomic number rule'

While exceptions to the octet rule are relatively rare for first row main group compounds, the *18–electron rule* is contravened, or is apparently not a factor, in many instances. Initially this all appears very confusing but in practice the observed configurations may be rationalized quite well.

The 18–electron rule states that 'the number of valence electrons of the metal together with the number of electrons provided by the ligands equals the number of electrons possessed by the next noble gas'. In other words

Table 6.4. Some examples of 18-electron complexes across the periodic table.

5	6	7	8	9	10	11	12
$[V(CO)_6]^-$	$[Cr(CO)_6]$	$[Mn(CO)_6]^+$	$[Fe(CN)_6]^{4-}$	$[CoF_6]^{3-}$	$[Ni(PF_3)_4]$	$[Cu(CN)_4]^{3-}$	$[Zn(CN)_4]^{2-}$
$[Nb(CN)_8]^-$	$[Mo(CN)_8]^{4-}$	$[Tc(CO)_3Cp]$	$[Ru(NH_3)_6]^{2+}$	$[RhCl_6]^{3-}$	$[Pd(PPh_3)_4]$	$[Ag(SCN)_4]^{3-}$	$[Cd(NH_3)_4]^{2+}$
$[Ta(CO)_6]^-$	$[W(CO)_5]^{2-}$	$[Re(CO)_5]^-$	$[Os(CO)_5]$	$[IrH_3(PPh_3)_3]$	$[Pt(PF_3)_4]$	$[AuF_6]^-$	$[Hg(NO_2)_4]^{2-}$

complexes attain kinetic stability when all the metal *nd*, $(n+1)s$, and $(n+1)p$ orbitals are filled with electrons, 18 of which are required for this purpose.

Consider the structures and stoichiometries of the three carbonyl complexes so far encountered. Why are these particular stoichiometries adopted? What is wrong with $[Fe(CO)_6]$? After all, the anticipated geometry would be octahedral which is very common, indeed $[Cr(CO)_6]$ (Fig. 6.28) is octahedral and is a very stable molecule. Chromium in $[Cr(CO)_6]$ is in the zero oxidation state and therefore has the configuration d^6. The six carbonyls donate a total of 12 electrons to the metal, making 18 in total. Similar calculations for both $[Fe(CO)_5]$ and $[Ni(CO)_4]$ show that these are also 18 electron species. There is a clear relationship between the d^n configuration and the coordination number for this series. For every two extra electrons in the d^n configuration, the coordination number drops by one (provided the ligands are simple two electron donors).

If a metal in a complex were to exceed the 18-electron count, there would have to be an extra low lying orbital that could accept electrons from an extra ligand. Part of the reason that the 18-electron rule is adhered to so well for complexes of π-acceptor ligands is that there are *normally* no low lying energy levels readily available for interaction with other ligands once the 18-electron count has been achieved. Generally, the next available energy levels are high lying and antibonding.

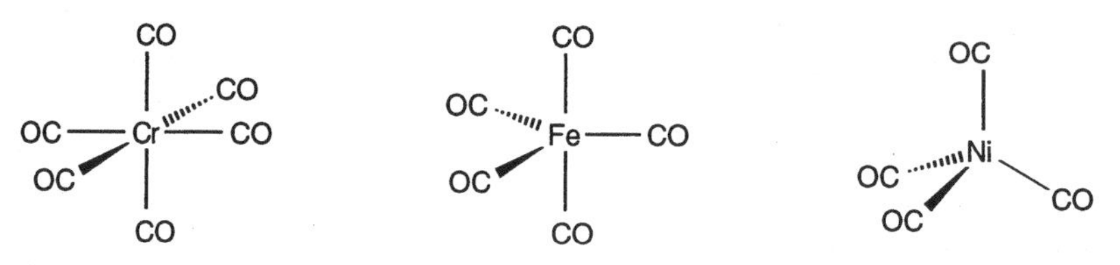

Fig. 6.28 The structures of three metal carbonyl complexes

6.11 The effect of π-interactions upon the valence e^- count

At this point the reader is encouraged to browse through the sections on *d*-block metals in one of the larger texts referred to in the reading list. The observant reader will discover some octahedral complexes for which the 18-electron count is exceeded. How can this be? It turns out that the degenerate e_g set of two of antibonding orbitals (Fig. 6.9) are only weakly antibonding in some cases. That is, they are quite low in energy. Under these circumstances, the e_g orbitals may be occupied. This means that the 18-electron count can be exceeded, and that some compounds possess up to 22 valence electrons. So, whereas main group elements from the second and subsequent periods can expand their octet by using appropriate *d* orbitals, octahedral transition metal compounds can sometimes expand beyond the 18-electron count by using low-lying antibonding orbitals. Discussion is limited here to octahedral complexes, but the same argument applies to other geometries. There are three (overlapping) situations:

- those with a small Δ_o because the ligands make relatively weak σ-bonds ('weak σ-effect ligands') and for which there are few or no π-effects
- those with a large Δ_o because the ligands make strong σ-bonds ('strong σ-effect ligands') and for which there are little or no π-interactions
- those with a large Δ_o because the ligands make strong σ-bonds and for which there are strong π-acceptor interactions

The central sections of the energy level diagrams for these three cases are shown in Fig. 6.29.

Weak σ-donors, weak π-effect donors

If the σ-bonds are weak, this means the six bonding orbitals are not very low in energy, neither are the six antibonding orbitals very high in energy. Since the e_g orbitals (two of the antibonding orbitals) are not very high in energy, they cannot be far away from the nonbonding t_{2g} orbitals, and so Δ_o must be small.

Since Δ_o is small, there is only a small energy penalty for putting electrons into even the weakly antibonding e_g level. The additional bond energy gained on forming six M—L bonds rather than just five or four M—L bonds can

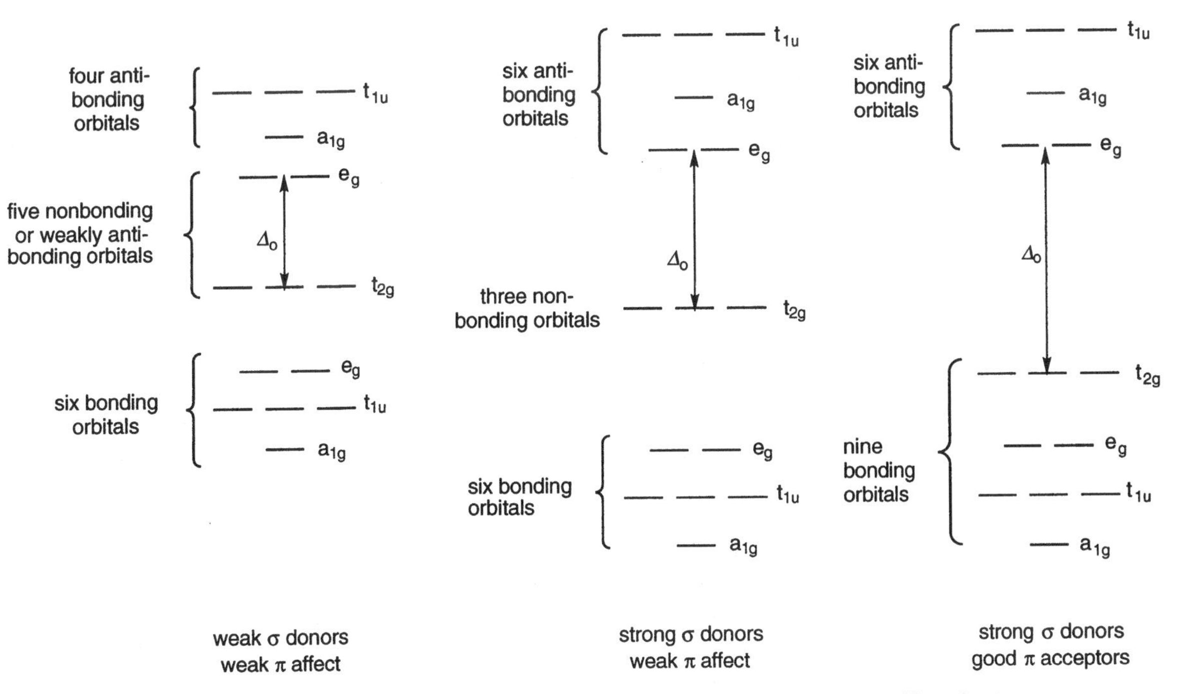

Fig. 6.29. The central sections of the energy level diagrams for the three types of ligand set.

provide this extra energy. In effect, there are six low energy molecular orbitals which are filled by ligand electrons. There are then five intermediate energy levels which may or may not have electrons according to the particular circumstances of the compound. There can therefore be anywhere from 12 to 22 valence electrons in this type of complex. For these compounds the specific number 18 has little advantage. Examples of 12- to 22-electron complexes within this class are known.

Strong σ-donors, weak π-effect donors

If the σ-bonds are strong, then the six bonding orbitals are low in energy and the six antibonding σ-orbitals must be high in energy. Since the e_g orbitals (two of the antibonding orbitals) are relatively high in energy, they are far away from the nonbonding t_{2g} orbitals, and so Δ_o is large.

Since Δ_o is large because of the strong σ–donor effects of the ligands, there are now only three intermediate energy levels which can accommodate electrons. The e_g level is now out of reach and cannot hold electrons but the t_{2g} set can hold electrons if necessary. There are still the six strongly bonding orbitals holding the ligand electrons. In these cases the complexes possess anywhere between 12 and 18 metal valence electrons (Table 6.6).

Strong σ-donors, strong π-effect donors

The final case is where the ligands are both strong σ-donors and π-acceptors. The strong σ-donation makes the e_g levels very high in energy, too high for occupation. The π-acceptor properties of the ligands makes the t_{2g} set very low in energy (Section 6.5) and it is now overwhelmingly desirable that these should be filled. There are therefore no intermediate energy levels; every level is either strongly bonding or antibonding. There are a total of nine low energy molecular orbitals to be filled. This means that an octahedral geometry is very common for d^6 metal complexes in which the ligands are π acceptors (such as $[Cr(CO)_6]$ or $[Co(CN)_6]^{3-}$). An eighteen electron configuration is therefore very stable for strong crystal field effect ligands and it is to these types of complex that the 18-electron rule applies in particular.

6.12 Metal—metal bonding

Calomel is a compound of mercury known to Indian chemists from very early times. Although not appreciated at the time, calomel, $[Hg_2Cl_2]$, is linear and contains a direct metal–metal bond. Many other compounds with metal–metal interactions are known and there is a lively discussion in the research literature about their nature.

One case which seems relatively straightforward is that of the Group 7 carbonyl dimers $[M_2(CO)_{10}]$ (M = Mn, Re). The solid state structure of $[Re_2(CO)_{10}]$ is shown in Fig. 6.30 and shows that each metal is octahedrally coordinated to five carbonyl groups with the sixth ligand site in each case being taken up by the other metal atom. In effect, each $M(CO)_5$ unit functions as a ligand to its partner. The arrangement is such that the carbonyl groups are staggered, rather as the lowest energy arrangement of ethane is with the

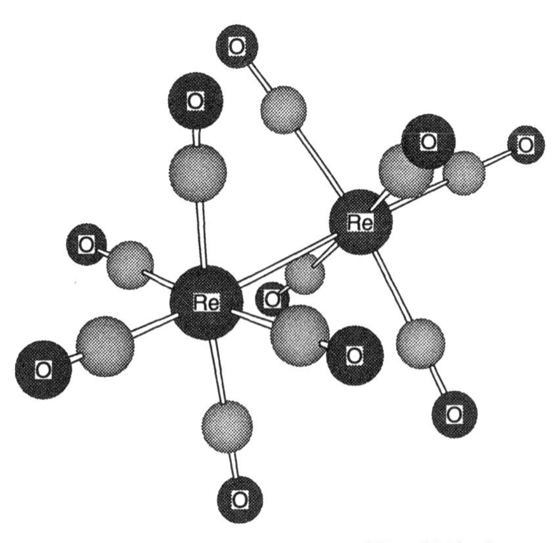

Fig. 6.30. Solid state structure of $[Re_2(CO)_{10}]$.

hydrogen atoms in a staggered arrangement. Another example of a structurally related metal—metal bonded species is the complex anion $[Co_2(CN)_{10}]^{6-}$.

Fig. 6.31. The unusual nature of $[Re_2Cl_8]^{2-}$ is recognized by this issue of a Russian stamp in 1968. It was issued to commemorate the 50th anniversary of the Kurnakov Institute for General and Inorganic Chemistry and portrays the dimeric ion $[Re_2X_8]^{2-}$; Scott catalogue #3507, Scott Publishing Co., Sidney, Ohio, USA.

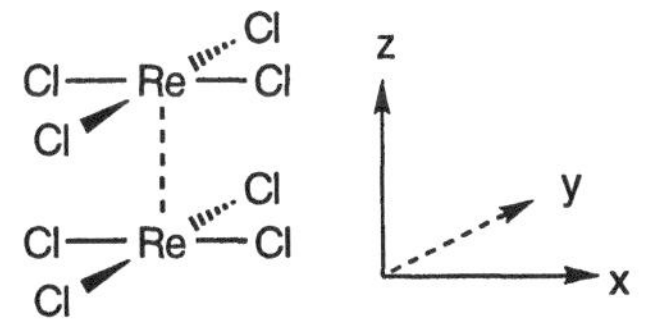

The dianionic complex $[Re_2Cl_8]^{2-}$ (Figs 6.31 and 6.32) is another interesting compound. Formally, it consists of two d^4 $ReCl_4$ groups containing Re(III) with the two metals separated by 224 pm. In this case the chloride groups are arranged in an eclipsed configuration. The effect is for the eight chloride groups to form an almost perfect cube with the two metal atoms slightly offset from two opposite faces of the cube. But why are the chlorides eclipsed? From a steric and electrostatic point of view, the staggered conformation would seem to be preferred. The answer lies in the bonding. There are several ways to provide a simple pictorial view of the bonding and one useful way to picture the bonding is as follows. The dimer consists of two d^4 $[ReCl_4]^-$ groups and so it is appropriate to consider the bonding in this fragment first. To a first approximation, the $[ReCl_4]^-$ group is square planar and therefore dsp^2 hybridized.

The hybridization scheme involves the s, p_x, p_y, and $d_{x^2-y^2}$ orbitals. The next step is to present the two $[ReCl_4]^-$ groups to each other. There are two orbitals suitable for σ bonding: the d_{z^2} and p_z orbitals. These two orbitals are hybridized together. The result is one hybrid orbital which is directed towards the second metal (and so towards an analogous hybrid orbital) and a second which is directed away from the second metal. This second hybrid orbital is empty and essentially nonbonding. The first hybrid orbital makes an overlap with the corresponding orbital on the second metal to make a M—M s bonding orbital and a M—M σ^* orbital.

This leaves the remaining three d orbitals. Two of these, the d_{xz} and d_{yz} are set to overlap with the corresponding orbitals on the second atom in a π sense. These two interactions are degenerate. There is also an associated degenerate pair of π^* orbitals. The final orbital, the d_{xy} orbital overlaps with the corresponding orbital on the second metal in such a way that all four lobes are involved (Fig. 6.33). This type of overlap is called a δ bond. There

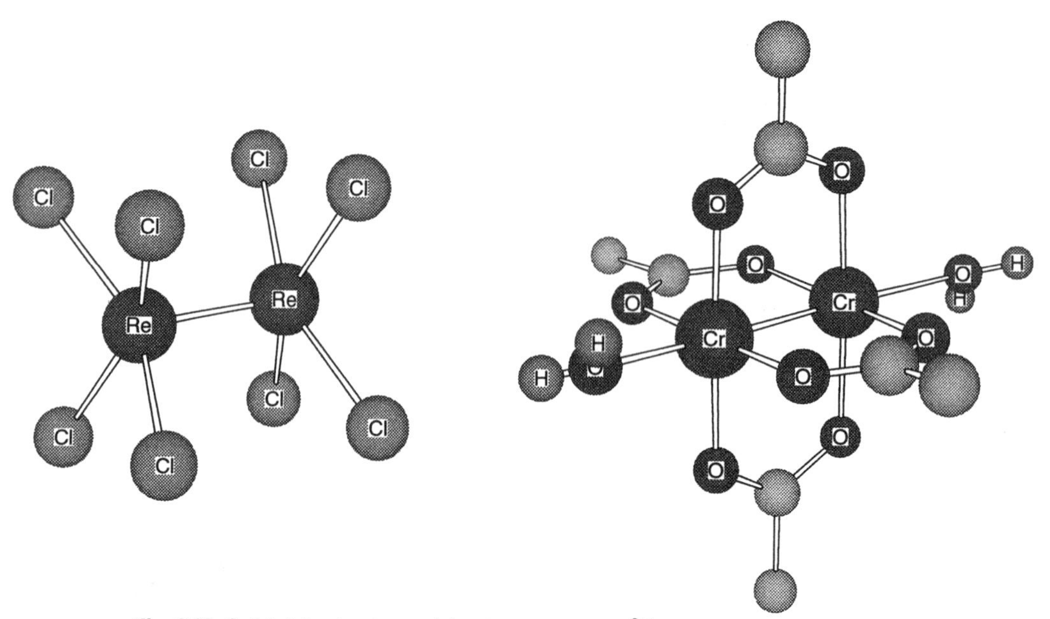

Fig. 6.32. Solid state structures of the dianion $[Re_2Cl_8]^{2-}$ and $[Cr_2(O_2CMe)_4(OH_2)_2]$.

is also an associated δ^* orbital. Calculations suggest that the δ bond is the weakest of these bonds, but it is this overlap which forces the chlorides to be eclipsed.

The origin of the name 'δ' lies in the fact that the *d* bond looks like a *d* orbital when viewed along the internuclear axis

Chromium acetate is another molecule with a quadruple metal–metal bond. Note that the metal, Cr(II) in this case, is also d^4. Again, (Fig. 6.33) the eight ligated atoms are eclipsed and the nature of the bonding is related to that of $[Re_2Cl_8]^{2-}$. In this case the unoccupied hybrid orbitals referred to above are available for binding to a ligand and one characteristic feature of chromium acetate is its ability to coordinate ligands on the sites *trans* to the M—M quadruple bond.

It is quite common for M—M bonds, whether multiple or single to be bridged by ligands such as halide or carbonyl (Fig. 6.34). For electron counting purposes, bridging carbonyl groups are regarded as donating one electron to each of the two metals.

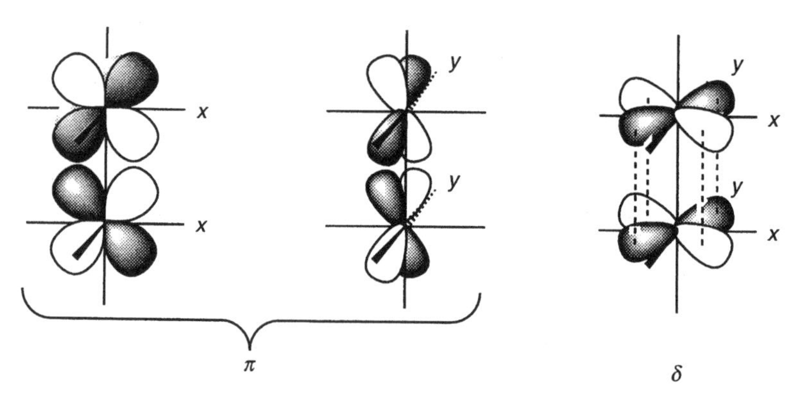

Fig. 6.33. The π and δ bonds in quadruple metal—metal bonded complexes.

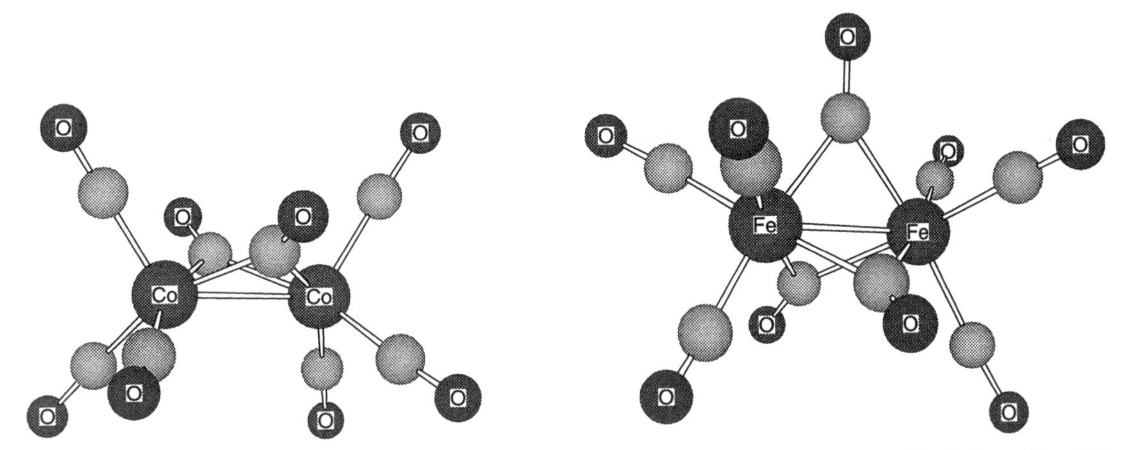

Fig. 6.34. Solid state structures of complexes with bridging ligands supporting M—M bonds: $[Co_2(CO)_8]$ (left) and $[Fe_2(CO)_9]$.

6.13 Exercises

1. The structures in Fig. 6.34 contain examples of metal complexes in which carbonyl groups bridge two metals. Find examples of complexes in which there are bridging ligands other than carbonyl. For each, describe a simple way of describing the nature of the bridging interaction.
2. Explain why there are relatively few metal complexes containing π-acceptor ligands that disobey the 18-electron rule.
3. Construct an energy level diagram similar to Fig. 6.2 but for square planar, tetrahedral, and trigonal bipyramidal geometries.

7 Consequences of *d*-orbital splitting

In Chapter 1 reference was made to special properties of transition metal complexes such as their colour and unusual magnetic properties. Armed with the material in subsequent chapters it is possible to provide explanations for these properties.

7.1 Spectroscopy and the measurement of Δ_o

A solution of Ti(III) in water consists of a solution of the complex ion $[Ti(OH_2)_6]^{3+}$, a d^1 complex. Its colour is violet. The advantage of looking at a d^1 complex is that there is just a single electron to consider. The UV–visible absorption spectrum is shown schematically in Fig. 7.1. There is a single peak in the visible part of the spectrum and its maximum around 20300 cm^{-1} lies in the yellow–green part of the spectrum so the colour observed is white light with green subtracted out, that is, violet. The colour is a consequence of the movement of electrons between the *d* levels which are split by the crystal field effects of the six water ligands.

In general, if a compound contains an electron in the lower of two energy levels, then irradiation of that compound with light whose energy is equal to the energy gap between the two levels will cause absorption of light and the promotion of the electron to the higher energy level. The new electronic configuration is called an *excited state*.

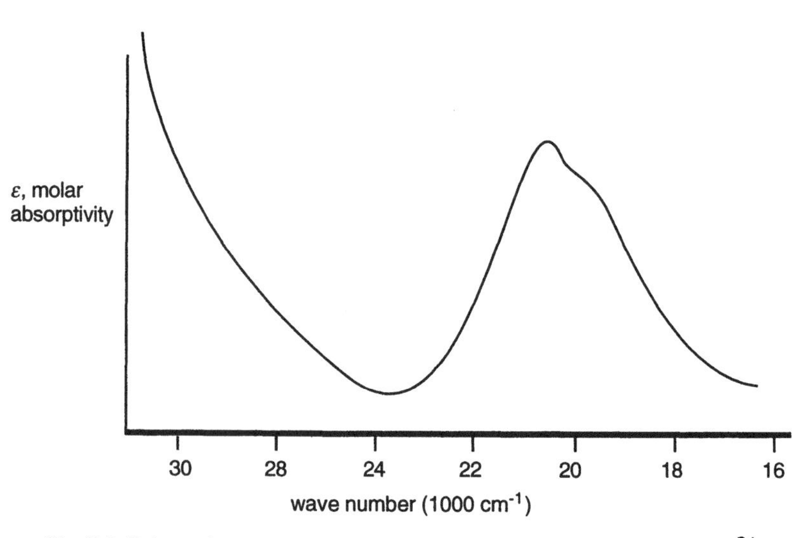

Fig. 7.1. Schematic representation of the UV/visible spectrum of $[Ti(OH_2)_6]^{3+}$.

The effect of shining white light onto the $[Ti(OH_2)_6]^{3+}$ ion is that some light whose energy is equal to the difference in energy between the t_{2g} and the e_g levels is absorbed. The energy of the light corresponds to the energy value of Δ_o and allows direct measurement of this quantity. Frequently, one of the easiest ways to measure values for Δ_o is to use UV/visible spectroscopy. In the case of $[Ti(OH_2)_6]^{3+}$, the light which is absorbed lies in the visible region, accounting for the compound not being colourless.

7.2 Selection rules for transition metal complexes

The absorption band in the electronic spectrum of $[Ti(OH_2)_6]^{3+}$ is actually rather weak. The value of the molar absorbance at the maximum is about 5 whereas a strong absorption (say, the colour of a solution of potassium permanganate) has a molar absorbance around 10^4–10^5. The problem is that while one can point to an energy level containing an electron and then point to another empty energy level at a higher energy, it does not necessarily follow that the transition will actually take place. Whether or not a transition will take place between two energy levels is governed by *selection rules*. Metal complexes are not the only species governed by selection rules, all compounds are so governed. If a transition obeys a given set of selection rules then that transition is said to be *allowed*. If a transition does not follow the rules, the transition is *forbidden*. There are two rules of importance in the context of *d*-block metal complexes.

Symmetry selection rules (Laporte rule)

A function which possesses precisely the same value at a point $(-x, -y, -z)$ as at (x, y, z) is said to possess a centre of symmetry. Under this definition functions representing *s* and *d* orbitals (Fig. 7.2) possess a centre of symmetry while that representing a *p* orbital does not. For a *p* orbital the absolute magnitude of the function at $(-x, -y, -z)$ is the same as at (x, y, z), but the sign is different. Follow the dotted line through the origin for each orbital. The sign of the wave function stays the same for the *s* orbital and also for the *d* orbital. On the other hand the sign of the wave function changes on crossing the nucleus for the *p* orbital.

Orbitals which possess a centre of symmetry when in a molecule that itself possesses a centre of symmetry are said to be *gerade* (g). Those which do not are called *ungerade* (u). Hence the subscript g in the terms t_{2g} and e_g used for labelling the split *d* orbitals in an octahedral field. Since the tetrahedron does

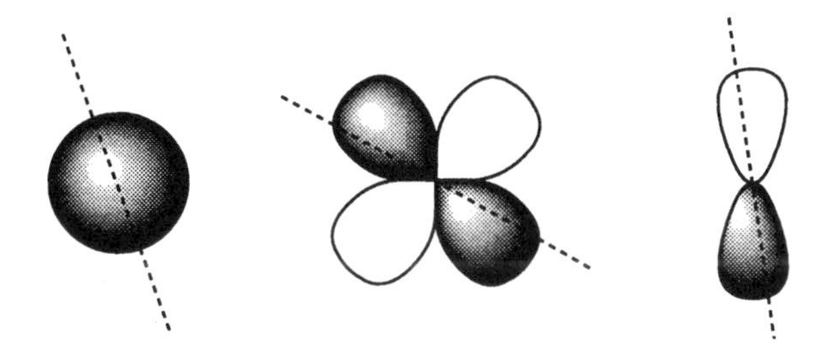

Fig. 7.2. The *s* and *d* orbitals possess a centre of symmetry, the *p* orbital does not.

not have a centre of symmetry, it is not appropriate to use the labels g and u and so the *d* orbitals when split in a tetrahedral crystal field are labelled e and t_2.

In a molecule that possesses a centre of symmetry, such as an octahedral complex ML_6, for a transition to occur, the electron must move *either* from a *gerade* orbital to an *ungerade* orbital *or* from an *ungerade* orbital to a *gerade* orbital. Since all the *d* orbitals are gerade, this requirement is not met for ML_6 complexes so that *d–d* transitions are forbidden in octahedral complexes. This rule could be stated the other way round. Transitions involving gerade→gerade transitions ($d \rightarrow d$) are disallowed, as are ungerade→ungerade transitions ($p \rightarrow p$).

Tetrahedral complexes do not possess a centre of symmetry. The symmetry selection rule no longer applies in cases where there is no centre of symmetry, and so *d–d* transitions for tetrahedral complexes are not disallowed by the symmetry selection rule.

Spin selection rule

A second rule that must be followed is that promotion of an electron can only proceed if the orientation of the spin is the same in the excited state as in the ground state. The spin direction cannot change. This is illustrated in Fig. 7.3. In the transition on the left, the direction of the electron spin does not change. This transition is *allowed* by the spin selection rules. In the transitions on the centre and right, the spin changes direction. Neither transition is allowed by the spin selection rules.

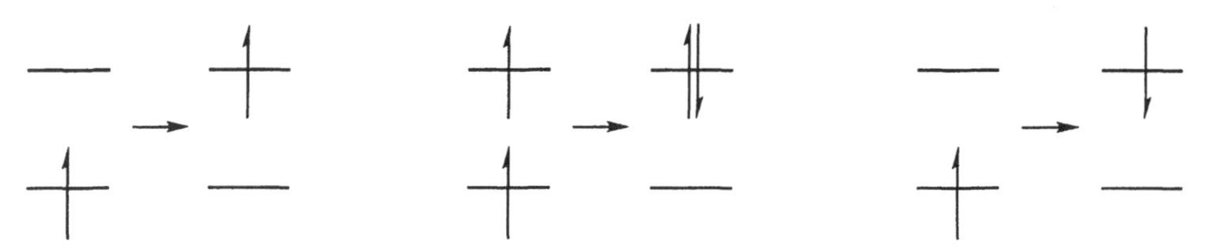

Fig 7.3. Examples of allowed (left) transition and forbidden transitions under the spin selection rule.

7.3 The low intensity of the *d–d* band in $[Ti(OH_2)_6]^{3+}$

The *d–d* transition in $[Ti(OH_2)_6]^{3+}$ is formally disallowed since the symmetry selection rule is broken. However there is still a weak absorption. Why is any absorption seen at all?

The large bandwidth of the absorption is due to molecular vibrations within the molecule which cause changes in the M—L bond lengths and so in the values of Δ_o. At any one instant, the nominally octahedral complex may be slightly asymmetric, not completely octahedral. This affects the value of Δ_o, which will be slightly different if the M—L distances are altered. These molecular vibrations take place rather more slowly than electronic transitions. Since at various times the molecule may be 'not quite' perfectly octahedral, it will 'not quite' have a centre of symmetry at all times. This allows the symmetry selection rule to 'not quite apply', allowing a weak signal to be visible.

While the symmetry selection rule is important for octahedral complexes, it does not apply for tetrahedral complexes. As a consequence, *d–d* bands in spectra of tetrahedral complexes are much stronger than those of octahedral complexes.

7.4 Charge transfer spectra

There is another class of electronic transition that can always occur in transition metal complexes in addition to *d–d* transitions. Transitions from one *d* orbital to another involve redistribution of electrons in orbitals that are basically pure metal orbitals. Another class of transitions involves the movement of electrons from orbitals that are essentially ligand in character to orbitals which are essentially metal in character (or vice versa). Since this involves the movement of an electron from one atom to another, the effect is the transfer of charge from one part of the molecule to another. Such transitions are therefore called *charge transfer transitions* and the bands in the UV–visible spectra assigned to such transitions are called *charge transfer bands*.

These bands are often extremely intense. Look again at the UV–visible spectrum of $[Ti(OH_2)_6]^{3+}$. The feature to the left represents the 'edge' of a charge transfer band in the ultraviolet region. They tend to be in the near UV but often appear in the visible region. The colour is so intense since the transitions break no selection rules of any kind. The bands might be 10^3 times stronger than *d–d* transition bands.

Familiar charge transfer colours are those associated with the deep red colour produced on addition of SCN^- to solutions containing Fe(III), and the deep purple colour of $KMnO_4$. The yellow brown colour of $[Fe(OH_2)_5(OH)]^{2+}$ is due to an OH to metal transition in the near UV which is so intense that the low energy edge absorbs significantly in the blue region of the spectrum. These transitions involve ligand → metal charge transfer. Other charge transfer transitions may involve metal → ligand charge transfers. These are displayed by complexes of π-acceptor ligands such as CO and might involve transfer of a metal electron into a CO π^* orbital.

7.5 Magnetism

Compounds with no unpaired electrons are diamagnetic while those with one or more unpaired electrons are paramagnetic. Electrons possess charge and mathematically can be treated as being in motion inside the atom. A moving electric charge constitutes an electric current. Electric currents set up magnetic fields and in that sense an electron inside a molecule may be regarded as a miniature bar magnet. The magnetic moment of this miniature bar magnet interacts with any externally applied magnetic field.

When an orbital contains two electrons, the consequence is that the effects of the electrons in that orbital cancel each other out. The atom taken as a whole will only show a 'resultant' magnetic moment if there are some unpaired electrons in the valence shell. When there are some unpaired electrons in the atom, the moments of the atoms will align themselves with

Paramagnetic complexes are drawn into a magnetic field

any external magnetic field. This behaviour is *paramagnetism*. Paramagnetic substances are attracted into magnetic fields, such as those supplied by a small electromagnet.

One Bohr magneton is the magnetic moment associated with a single electron in the Bohr hydrogen atom

Diamagnetic compounds are repelled from a magnetic field

The magnitude of this attraction is dependent on the number of unpaired electrons in that molecule. A paramagnetic atom or molecule is characterized by a number called the effective magnetic moment, μ_{eff}, which is usually expressed in units of Bohr magneton (μ_B).

All substances, whether containing unpaired electrons or not, are affected to a degree when placed in a magnetic field. This is because the magnetic field will distort the motion of the electrons within the atoms with the consequential formation of a small current, and hence a small induced magnetic field. A law in physics called Lenz's law states that such induced fields will always oppose the applied field that set them up in the first place. This causes the atoms or molecules to be repelled from an applied magnetic field. This behaviour is called *diamagnetism*. Diamagnetic effects are much weaker than paramagnetic effects. When there are some unpaired electrons in a compound, the paramagnetic effect swamps the diamagnetic effect and overall the molecule appears paramagnetic.

Paramagnetic effects originate from two sources. They arise from the spin angular momentum and the orbital angular momentum of the electrons. The latter is a function of whether the electron is in an *s*, *p*, *d*, or *f* orbital. The orbital contributions are generally suppressed by the ligands in complexes of first row elements. Orbital contributions become more important for complexes of second and third row elements.

There is no *simple* way to include the orbital effects, and it is common and often adequate to only consider spin effects. The effective magnetic moment (in μ_B units) is expressed by the *spin only* equation (Eqn 7.1) in which n is the number of unpaired electrons. The equation including the orbital contribution in addition is more complex.

$$\mu_{eff} = \sqrt{n(n+2)} \tag{7.1}$$

The practical consequence of these effects is that when weighed in the absence of an applied field and subsequently in the presence of a magnetic field (i.e., turn the magnet current on), *diamagnetic* substances show a *decrease* in recorded weight and *paramagnetic* substances show an *increase* in recorded weight. From the weight difference and with a knowledge of the stoichiometry of the compound, one can calculate the magnetic moment of the compound, and hence the number of unpaired electrons within it. Some examples of calculated and effective magnetic moments are included in Table 7.1.

There are a number of uses for the measurements of magnetic moments. The important one here is that usually one can determine whether a complex is high or low spin simply by doing one or two reasonably accurate weighings. For instance, a complex of Fe(III), d^5, can be low spin (one unpaired electron) or high spin (five unpaired electrons). The theoretical magnetic moments using the spin only formula are 1.73 or 5.92 μ_B

Table 7.1. Calculated and observed values of μ_{eff} for some first row complexes.

complex	n	μ_{eff} / μ_B	μ_{obs} / μ_B
$[Ti(OH_2)_6]^{3+}$	1	$\sqrt{1(1+2)} = 1.73$	1.75
$[V(OH_2)_6]^{3+}$	2	$\sqrt{2(2+2)} = 2.83$	2.80
$[Cr(OH_2)_6]^{3+}$	3	$\sqrt{3(3+2)} = 3.87$	3.88
$[Mn(OH_2)_6]^{3+}$	4	$\sqrt{4(4+2)} = 4.90$	4.93
$[Fe(OH_2)_6]^{3+}$	5	$\sqrt{5(5+2)} = 5.92$	5.40
$[Co(OH_2)_6]^{2+}$	3	$\sqrt{3(3+2)} = 3.87$	4.85
$[Ni(OH_2)_6]^{2+}$	2	$\sqrt{2(2+2)} = 2.83$	2.83
$[Cu(OH_2)_6]^{2+}$	1	$\sqrt{1(1+2)} = 1.73$	1.75

respectively. Even allowing for possible effects of orbital contributions, the simple weighing should be able to distinguish between the two.

It is usual to construct a dedicated balance to carry out such weighings. The differences in weight are not dramatic so it is important to exclude sources of drafts and moisture since both easily affect a sensitive balance. A schematic representation of one such balance called a Gouy balance is shown in Fig 7.4).

7.6 Factors affecting the size of Δ_o

Having discussed some ways in which values of Δ_o can be determined, it is necessary to examine some factors upon which the size of Δ_o depends. There are a number of reasons why it would be useful to know these factors. For

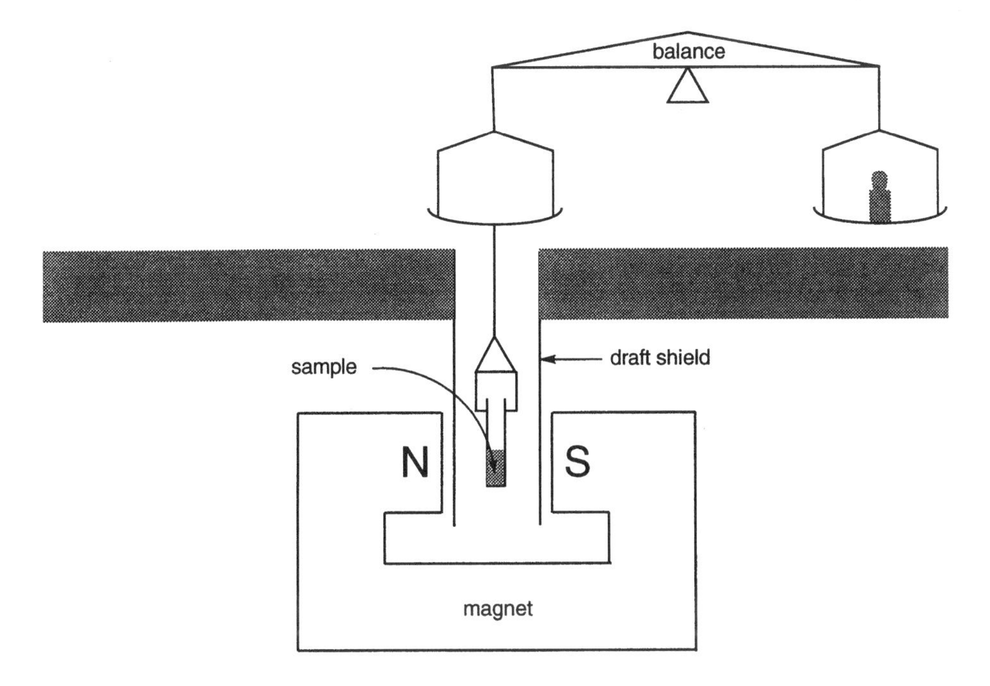

Fig. 7.4. Schematic representation of a Gouy balance. The magnet is an electromagnet. One would normally expect to carry out calibrations against samples of precisely known composition to compensate for systematic errors.

instance, if one could tune the size of Δ_o it should, in principle, be possible to predict whether a complex will be high or low spin and, or to predict its colour. Whether a complex is high or low spin is likely to have important consequences for the reactivity of that complex.

The following factors are illustrated using values of Δ_o, but the same arguments are applicable to Δ_t values and orbital splittings in other geometries.

Oxidation state of the metal

The greater the charge on the metal, the greater is Δ_o. If the charge is large, then this should cause the ligands to be attracted closer towards the metal, so interacting more strongly with the *d*-orbitals. As examples, for the first transition metal series Δ_o values for $[M(OH_2)_6]^{2+}$ (about 10000 cm^{-1}) complexes are about half those for the corresponding $[M(H_2O)_6]^{3+}$ complexes (about 20000cm^{-1}). This means that M(III) complexes are far more likely to be low spin whereas corresponding M(II) complexes might be high spin. As examples, the d^6 complexes $[Fe(NH_3)_6]^{2+}$ (μ_{obs} = 4.9 μ_B) and $[Co(NH_3)_6]^{3+}$ (μ_{obs} = 0.0 μ_B) are respectively high and low spin.

Transition series of the metal

Within a transition metal group, Δ_o increases by 20–50% on going from the first series metal to the second, and by another 20–50% on going from the second series metal to the third. This is a consequence of the increase in size of the *d* orbitals ($3d < 4d < 5d$). Bigger *d* orbitals interact more with the ligands. This is shown quite clearly in the following series. Note also that larger *d* orbitals (4*d*, 5*d*) have lower pairing energies, *P*. This also favours low spin configurations.

Table 7.2. Values of Δ_o for the complexes $[M(NH_3)_6]^{3+}$ (M = Co, Rh, Ir).

complex	$[Co(NH_3)_6]^{3+}$	$[Rh(NH_3)_6]^{3+}$	$[Ir(NH_3)_6]^{3+}$
Δ_o (cm^{-1})	22900	34100	41000

The ligand

Consider Table 7.3. It shows some values of Δ_o for a number of octahedral complexes. At first sight there might seem little trend in any of the numbers. However a closer inspection will reveal that the value of Δ_o increases across the table. Since this is so for a number of metals, one can conclude that the value of Δ_o for CN^- is always greater than the value for en, which in turn is greater than the value for NH_3, and so on. If data were included for other metals with the same ligand sets, there would be nothing to upset this conclusion. In fact, by making a great number of measurements for a large number of ligands, one can arrange ligands in order of decreasing Δ_o, that is in order of decreasing crystal field strength. Such a list is called the *spectrochemical series*.

Table 7.3. Value of ligand field splitting parameter Δ_o for some octahedral complexes The units of Δ_o are in cm^{-1}. For comparison 10000 cm^{-1} = 119.7 kJ mol^{-1}. The values in **bold** type indicate that value is for a low spin complex.

ion (d^n)	ligand set				
	$6Cl^-$	$6H_2O$	$6NH_3$	3en	$6CN^-$
Cr^{3+} (d^3)	13700	17400	21500	21900	26600
Mn^{2+} (d^5)	7500	8500	-	10100	30000
Fe^{3+} (d^5)	11000	14300	-	-	**35000**
Fe^{2+} (d^6)	-	10400	-	-	**32800**
Co^{3+} (d^6)	-	**20700**	**22900**	**23200**	**34800**
Rh^{3+} (d^6)	**20400**	**27000**	**34000**	**34600**	**45500**
Ni^{2+} (d^8)	7500	8500	10800	11500	-

The spectrochemical series:

$$CN^- \approx CO \approx C_2H_4 > PR_3 > NO_2^- \approx \text{phen} > \text{bpy} > SO_3^{2-} > \text{en} \approx \text{py} \approx NH_3 > \text{edta}^{4-} > NCS^- > H_2O > C_2O_4^{2-} > ONO_2^- > OSO_3^{2-} > OH^- \approx ONO^- > F^- > Cl^- \approx SCN^- > Br^- > I^-$$

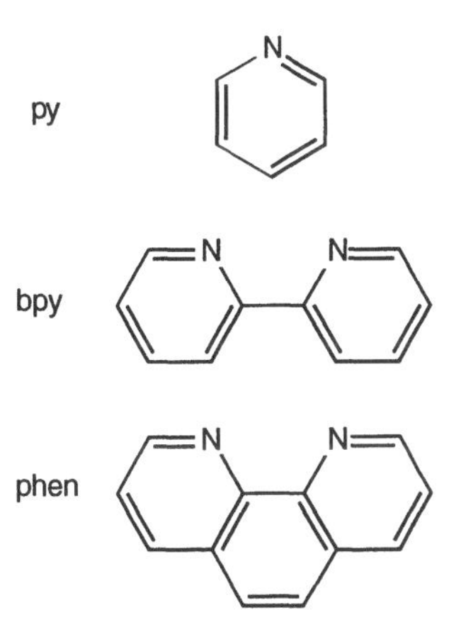

The ordering for the halides is in agreement with the order expected when one considers electrostatic effects. The iodide ligand is much bigger than F^- meaning that the charge representing the I^- will be further away, meaning less interaction with the *d* orbitals. Nitrogen donors have a greater effect than oxygen donors. The ligands with the strongest effect (CN^-, CO, etc.) are members of the π acceptor class. These show very strong covalent interactions with the metal.

Those ligands at the low Δ_o end of the series are known as *weak field* ligands, and those at the high Δ_o end as *strong field* ligands.

The spectrochemical series is very useful in predicting magnetic behaviour (that is, the electron configuration) of complexes. Thus, octahedral complexes of Fe^{2+} (d^6) are expected to be high spin for L = NH_3 and weaker field ligands, but low spin for L = bipy and stronger field ligands. Similarly Mn^{2+} (d^5) should be low spin only for CN^- and stronger ligands and Co^{3+} (d^6) should be high spin only with F^- and weaker ligands.

7.7 Distorted complexes: the Jahn–Teller effect

It is a fact that complexes which one might at first sight expect to be perfectly octahedral are distorted. Such distortions may be evident from the spectroscopic properties of the complex, or perhaps through an *X*-ray crystallographic study. Parts of the results for two such crystallographic studies are shown in Fig. 7.5. The diagrams show the structures of hexaaquo metal ions present in the salts $[NH_4]_2[Ni(OH_2)_6][SO_4]_2$ and $[Cu(OH_2)_6][ClO_4]_2$. In the first, the $[Ni(OH_2)_6]^{2+}$ ion is very symmetrical with all the O—Ni—O bond angles close to 90° and with all the Ni—O bond lengths about 207 pm. On the other hand, the corresponding $[Cu(OH_2)_6]^{2+}$ ion shows marked distortions. As for the nickel case, all the O—Cu—O bond angles are close to 90°, but now two mutually *trans* bond lengths are

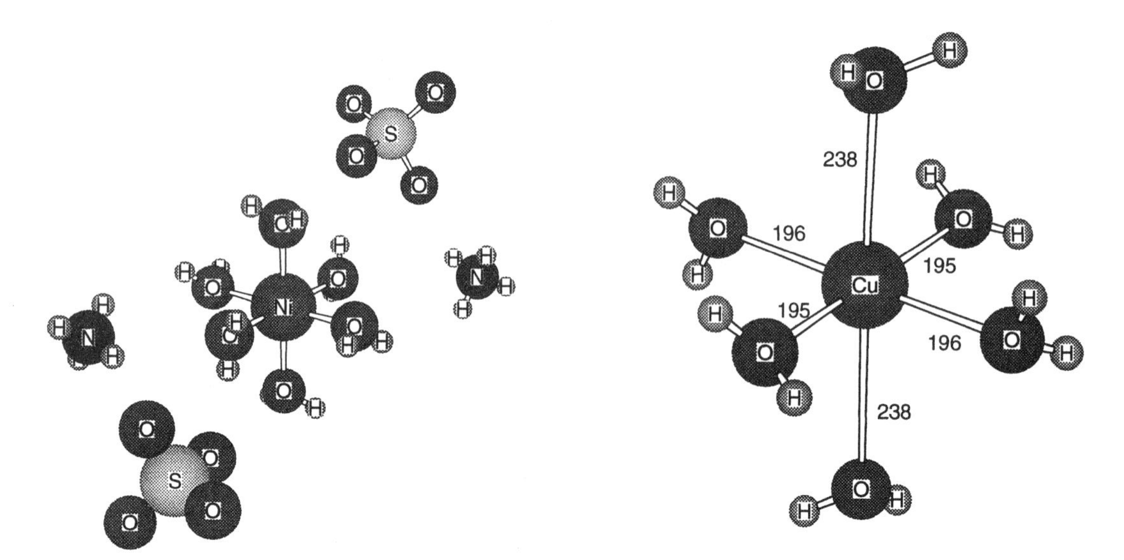

Fig. 7.5. Parts of the solid state crystal structures of the complex ions $[NH_4]_2[Ni(OH_2)_6][SO_4]_2$ (left) and $[Cu(OH_2)_6][ClO_4]_2$.

considerably longer than the remaining four. Since the ion still retains a four-fold axis of symmetry, the distortion is referred to as a *tetragonal* distortion.

These are not isolated examples. Unless there are extenuating circumstances, and making due allowances for cases in which not all ligating groups are the same, octahedral Ni(II) complexes tend to be undistorted, whereas Cu(II) complexes tend to show distortions. These distortions in Cu(II) complexes are often a consequence of the *Jahn–Teller effect*.

The distortion of Fig. 7.5 is a lengthening of two *trans* ligands and this seems to be the most common type of tetragonal distortion. Sometimes the distortion is in the opposite direction, in other words the two mutually *trans* ligands are closer to the metal than the remaining four ligands. For example, the X-ray crystal structure of K_2CuF_4 shows an array of fluoride ions in which each Cu(II) ion is surrounded by six fluoride ions, two *trans* fluorides at 195 pm and four at 208 pm. Curiously, the corresponding sodium salt, Na_2CuF_4, shows the opposite type of distortion, with two *trans* fluorides at 237 pm and the other four at 191 pm.

From an electronic point of view, the principle difference between the two M(II) ions in the structures of $(NH_4)_2[Ni(OH_2)_6](SO_4)_2$ and $[Cu(OH_2)_6](ClO_4)_2$ is that Ni(II) is d^8 and Cu(II) is d^9. The population of the d orbitals for Ni(II) is therefore $t_{2g}{}^6e_g{}^2$. The two e_g orbitals each contain one electron. The situation is more complex for the d^9 Cu(II) case. The population of the d orbitals is $t_{2g}{}^6e_g{}^3$ and the energy level diagram is derived from that of the d^8 configuration by adding an extra electron to the e_g level. The problem is this: should it be placed into the d_{z^2} orbital or into the $d_{x^2-y^2}$ orbital (Fig. 7.6)?

The effect of putting the ninth electron into the d_{z^2} orbital is to increase the amount of charge density along the z axis. In the crystal field approximation, this means that the point charges representing the ligands along the z axis cannot approach as closely as when there is just one electron in the d_{z^2}

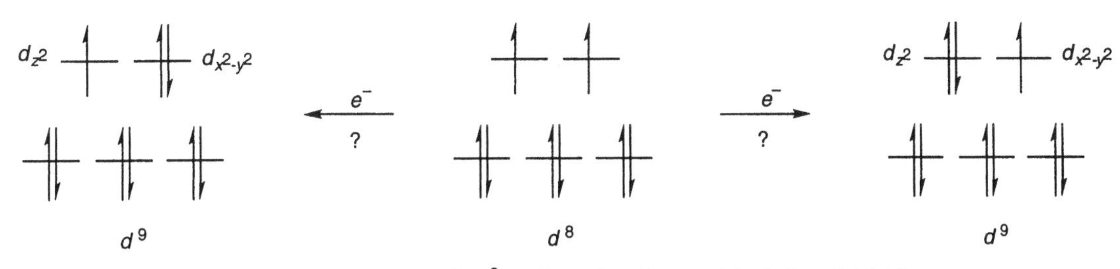

Fig. 7.6. The two possible d^9 configurations in an octahedral crystal field.

orbital. The bond lengths along the z axis will be longer than those along the x and y axes. Effectively, this is elongation along the z axis and the resulting distorted octahedron is a *stretched* octahedron. This type of distortion is called tetragonal since its symmetry is fourfold and is a *tetragonal elongation*. Since the z axis ligands are further away from the metal, the increase in the energy of the d_{z^2} orbital caused by approach of the z axis ligands is less. The energies of the d_{z^2} and $d_{z^2-y^2}$ orbitals are different, that is, the degeneracy of these orbitals is *lifted*.

The effect of longer z axis M—L bonds parallels the concept of partial z axis charge removal (Section 5.5) in deriving the crystal field diagram for square planar complexes. The energy level diagram looks the same (qualitatively) and is shown in the right of Fig 7.7. Note that there is a lesser splitting of the t_{2g} set of orbitals. The two orbitals with components along the z axis (the d_{xz} and d_{yz} orbitals) interact less with the z axis charges when the z axis bond lengths are longer and so appear in the energy level diagram below the energy of the d_{xy} orbital. The nine electrons are placed as shown and the only orbital that is not filled is the $d_{x^2-y^2}$ orbital.

If the ninth electron goes into the $d_{x^2-y^2}$ orbital, the result is that the ligands along the x and y axes are moved outwards a little. The orbitals with components in the x–y plane ($d_{x^2-y^2}$, d_{xy}) are stabilized somewhat relative to the other orbitals. The $d_{x^2-y^2}$ orbital drops by more than the d_{xy} orbital since it points more directly at the ligands. This results in the four M—L bonds along the x and y axes being longer than the corresponding M—L bonds along the z axis. This distortion is still tetragonal but is now a *tetragonal compression*.

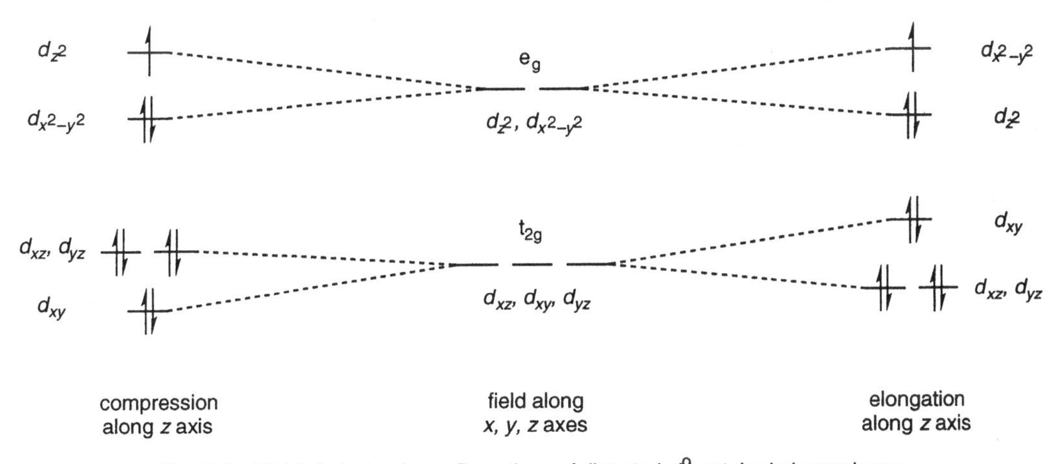

Fig. 7.6. d Orbital electronic configurations of distorted d^9 octahedral complexes.

This situation is shown in the left of Fig. 7.6.

In either case the degeneracy of the e_g set is lifted. If the electron moves into the d_{z^2} orbital, the d_{z^2} is stabilized relative to the $d_{x^2-y^2}$ orbital and the degeneracy is no longer present. If the electron goes into the $d_{x^2-y^2}$ orbital, the $d_{x^2-y^2}$ orbital is stabilized relative to the d_{z^2} orbital and again the degeneracy is lifted. In both cases structural distortions arise.

The reason for the ninth electron producing the Jahn–Teller distortion is effectively because the ninth electron goes into a set of degenerate orbitals in such a way as to produce an asymmetric electron population (two in one orbital and one in the other). In principle, a Jahn–Teller distortion will arise for *any* electronic configuration where such an asymmetry is possible. In practice, the distortion may not be easy to see.

A distortion is not seen for Ni(II) because the population of the e_g orbitals is symmetrical with one electron in each. Inspection of the energy level diagrams described in Chapter 5 for the d^1–d^{10} configurations of octahedral and tetrahedral complexes allows a prediction of which configurations might show distortions. For instance, in the d^4 high spin octahedral case, the fourth electron can go into either one of the e_g orbitals and a Jahn–Teller distortion will arise. This is well illustrated for the compound Cr_2F_5. This is a *mixed valence* compound of Cr(II) and Cr(III). The structure is based on a distorted cubic close packed array of F^- ions. Two-fifths of the available octahedral holes are filled by Cr^{2+} or Cr^{3+} ions in a symmetrical fashion. The Cr(III) ions are d^3, so the t_{2g} set of orbitals each contain a single electron. The configuration is symmetric and the holes in which the Cr(III) ions reside are perfectly symmetric. The Cr(II) ions are d^4 and so the octahedral holes in which the Cr(II) ions reside are distorted. The distortion is a tetragonal elongation. The two further F^- ligands lie at 260 pm from the metal and the four closer at 200 pm. Most Cr(II) structures show some distortion of this kind.

Distortions can also occur when the orbital occupation degeneracy occurs in the t_{2g} set of orbitals. In these cases, because the t_{2g} orbitals do not point directly at the ligands, the distortion is *smaller*. Sometimes, the effects are so small as to be undetectable. This could be the case for d^1 or d^2 configurations.

Now examine again the UV–visible spectrum of $[Ti(OH_2)_6]^{3+}$ (Fig. 7.1). The Ti(III) ion is d^1 and so prone to a Jahn–Teller distortion. In fact, a compression is exhibited and the electronic configuration of the d^1 case is shown in Fig. 7.8 after distortion. The single electron resides in the d_{xy} orbital. Now, upon promotion to the higher d levels (originally the e_g level), it can go into either of the two orbitals originally derived from the e_g set. The point is that different energies are required for each of these transitions. The energies are not very different. They are sufficiently similar that the bands overlap; the only evidence for the two bands is the shoulder. One could 'deconvolute' the spectrum mathematically and demonstrate that the spectrum really does consist of two overlapping peaks.

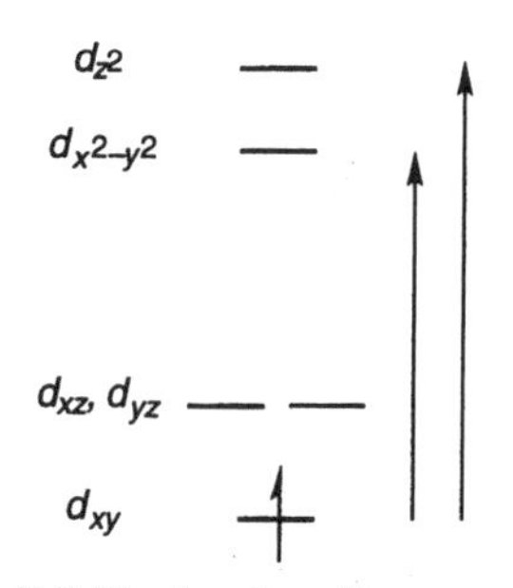

Fig. 7.8. The two transitions expected for a d^1 metal complex.

7.8 Periodic consequences of crystal field stabilization energies

The energy required for the formation of metal cations in the gas phase (the ionization enthalpy) follows a fairly smooth periodic trend from left to right across the periodic table (Fig. 7.9) with a slow increase as the effective nuclear charge of the metal increases. There are some notable discontinuities in the graph however. The plots for both the second and third ionization enthalpies both show a drop at the $d^6 \rightarrow d^5$ transition (the third ionization enthalpy for iron and the second for manganese). This is a consequence of interelectron repulsion between paired *d* electrons in a filled orbital making it easier to remove that sixth electron. However the periodic trends for metals contained within complexes are less smooth. This is partly because of crystal field effects.

The stabilization of complexes by crystal field stabilization effects affects the structure (and hence the reactivity) of many compounds. Recall from Section 5.4 that values for Δ_t are four-ninths the corresponding Δ_o values. Using these values one can plot the graph in Fig. 7.10 showing the CFSEs on a single graph using a Δ_o energy scale. This shows that by far the largest crystal field effects are shown for low spin d^5, d^6, and d^7 ions. Comparatively speaking the crystal field effects for all tetrahedral configurations are small. There is no net CFSE for the d^5 tetrahedral and high spin octahedral configurations.

The ionic radii of M^{2+} and M^{3+} ions are plotted for high and low spin environments in Fig. 7.11. Note the close similarity of the shape of this graph to the shape of the high and low spin CFSE graph in Fig. 7.9.

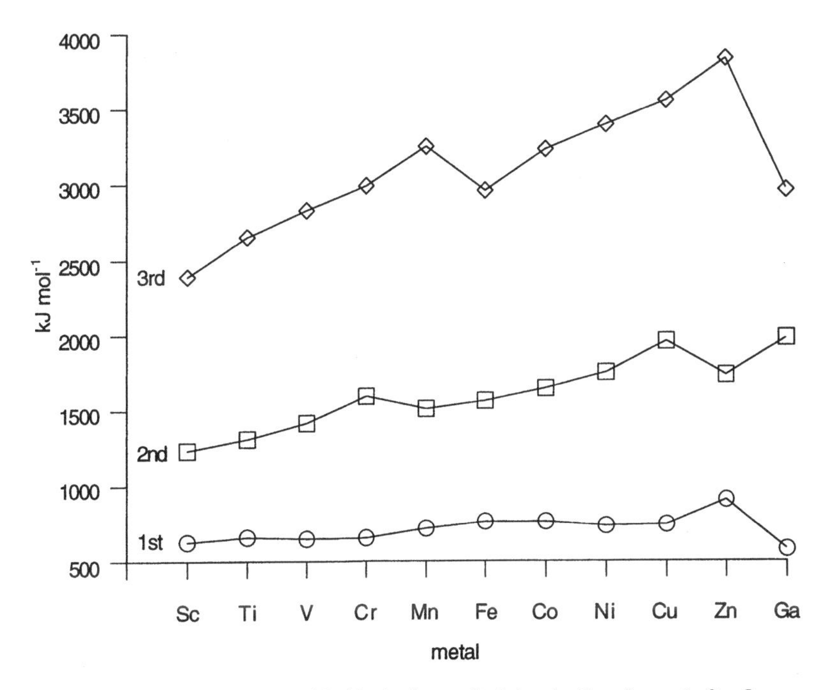

Fig. 7.9. Plot of 1st, 2nd, and 3rd ionization enthalpies for the elements Sc–Ga.

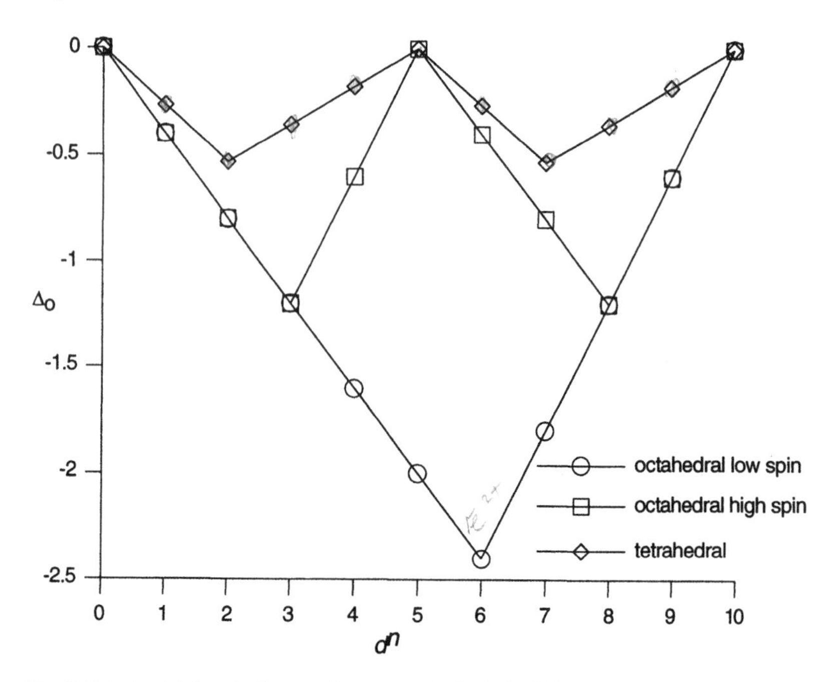

Fig. 7.10. A plot (excluding pairing energies) of CFSEs against the number of *d* electrons in a complex. Tetrahedral CFSE energies Δ_t are converted to units of Δ_o.

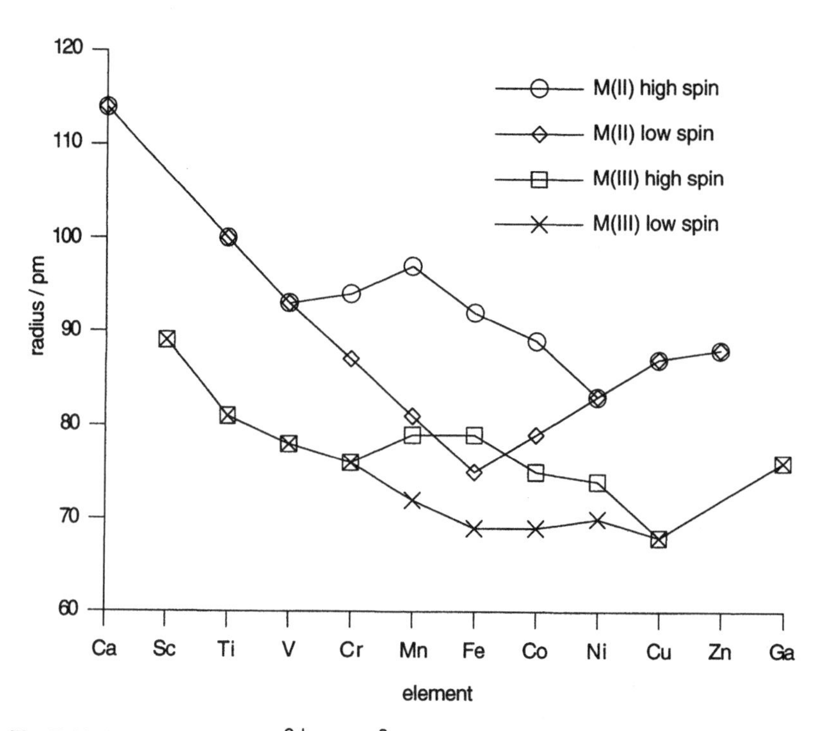

Fig. 7.11. Ionic radii for the M^{2+} and M^{3+} ions of the first transition series in high and low spin octahedral complexes.

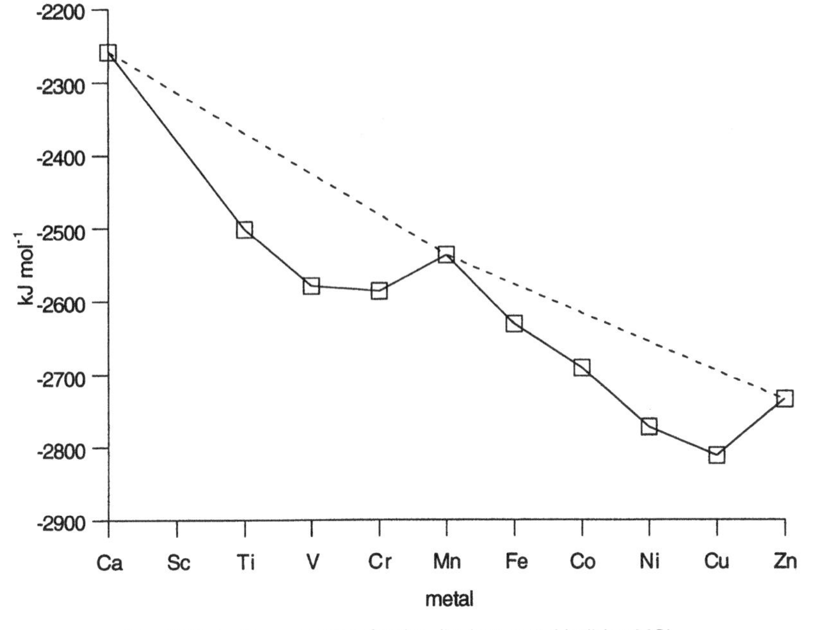

Fig. 7.12. Lattice energies for the divalent metal halides MCl_2.

The effect is particularly clear for M(II) ions where the minimum ionic radius for low spin ions is at Fe(II) and this corresponds exactly to the M(II) ion with the greatest CFSE. The two minima for the high spin ions at V(II) and Ni(II) also correspond exactly to the minima for the high spin CFSEs in Fig 7.10. The overall trend for ionic radii is for larger ions at the left of the periodic table and smaller ions to the right. This is because of the effective nuclear charge. In each case where there is a CFSE contribution, however, the ion appears even smaller than might have been expected. This is because the ions themselves might be regarded as 'lumpy' when in a crystal field environment. Consider the d^3 low spin case. Each of the three t_{2g} orbitals is occupied by one electron. These orbitals do not point directly at the ligand atoms, effectively allowing the ligands to approach closer along the ligand axes and making the size of the metal ion seem smaller along these axes (since the measurment of the ion radius is governed by the average M—L distance, by definition). Since the t_{2g} orbitals are directed into the regions between the ligands, the effective radius might be expected to be larger along those directions.

One effect of smaller ions is that the ligands approach closer to the metal ion. The closer approach of the ligands has consequences for the lattice energies and it is instructive to plot these on a graph (Fig. 7.12) for the high spin halides MCl_2 in which the metals are in an octahedral environment. Again the graph has a similar shape to Fig 7.10. This is a consequence of both extra lattice energy arising from the closer approach of the ligand and of the CFSE stabilization itself.

The enthalpies of hydration of the ions M^{2+} (Fig. 7.13) are very closely related to the enthalpy of formation of the complexes $[M(OH_2)_6]^{2+}$.

$$M^{2+}(g) + H_2O\ (excess) \rightarrow [M(OH_2)_6]^{2+}\ (aq)$$

All the complexes are high spin. The series of ions $Ca^{2+} - Zn^{2+}$ get smaller towards the right-hand side of the period as a consequence of increasing nuclear charge pulling the valence orbitals inwards. As a consequence a smooth increase in the exothermicity of the heat of hydration on crossing the period might have been anticipated. The actual trend (Fig. 7.13) is again very similar to the shape of the high spin CFSE plot in Fig 7.10. The deviation from the ideal curve is a consequence of the lumpiness of the M^{2+} described above allowing a closer approach of the H_2O ligands.

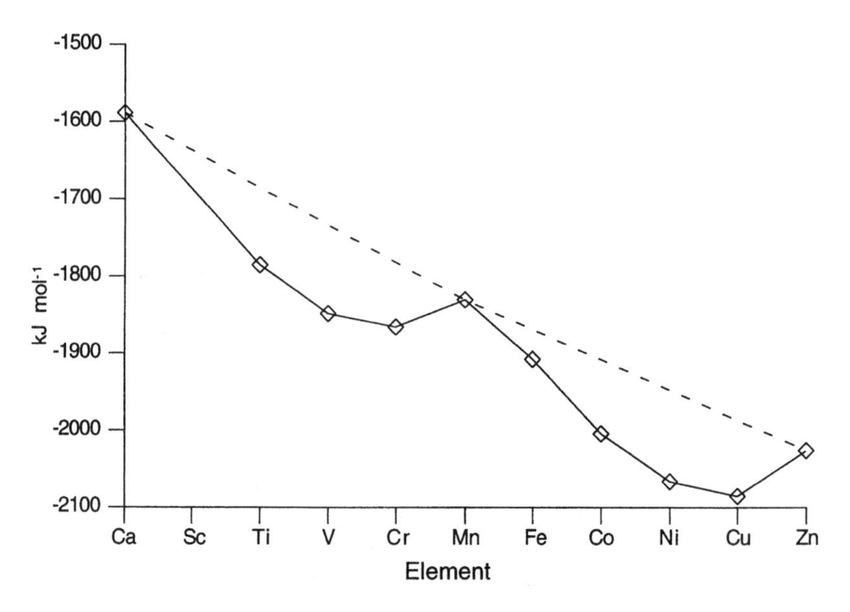

Fig. 7.13. Hydration energies of divalent metal ions. The solid line shows the observed data. The dotted line shows where the line should be if there were no CFSE effects.

7.9 Exercises

1. The following values of μ_{obs} are associated with each of the complexes listed. Determine which are high spin.

$[Fe(CN)_6]^{3-}$	2.3
$[CoF_6]^{3-}$	5.3
$[Co(NO_2)_6]^{4-}$	1.8
$K_2Mn(SO_4)_2.6H_2O$	5.9
$(NH_4)_2Co(SO_4)_2.6H_2O$	3.9

2. Examine some chemical data books and see if you can find other properties which when plotted suggest crystal field effects.
3. Make a list of d^n configurations that you would expect to display Jahn–Teller distortions in nominally octahedral and tetrahedral complexes.

8 Further reading

Atkins, P.W. and Beran, J.A. (1992). *General chemistry*, (2nd edn.). W.H. Freeman & Co., New York, USA.

Butler, I.S. and Harrod, J.F. (1989). *Inorganic chemistry – principles and applications*. Benjamin/Cummings Publishing Co., Inc., Redwood City, California, USA.

Cotton, F.A., Wilkinson, G., and Gauss, P.L. (1987). *Basic inorganic chemistry*, (2nd edn.). John Wiley and Sons, New York, USA.

Cotton, F.A. and Wilkinson, G. (1988). *Advanced inorganic chemistry*, (5th edn.). John Wiley and Sons, New York, USA.

DeKock, R.L. and Gray, H.B. (1980). *Chemical structure and bonding*. Benjamin/Cummings Publishing Co., Inc., Menlo Park, California, USA.

Greenwood, N.N. and Earnshaw, A. (1984). *The chemistry of the elements*. Routledge & Kegan Paul, London, UK.

Huheey, J.E., Keiter, E.A., and Keiter, R.L. (1993). *Inorganic chemistry – principles of structure and reactivity*. (4th edn.). Harper International., New York, USA.

Jolly, W.L. (1991). *Modern inorganic chemistry*. McGraw–Hill, Inc., New York, USA.

Mackay, K.M. and Mackay, R.A. (1989). *Introduction to modern inorganic chemistry*, (4th edn.). Blackie, London, UK.

Nicholls, D. (1974). *Complexes and first-row transition elements*. Macmillan Press, Basingstoke, UK.

Owen, S.M. and Brooker, A.T. (1991). *A guide to modern inorganic chemistry*. Longman Scientific and Technical, Harlow, UK.

Porterfield, W.W. (1984). *Inorganic chemistry – A unified approach*. Addison Wesley Publishing Co., Inc., Reading, Massachusetts, USA.

Purcell, K.F. and Kotz, J.C. (1985). *Inorganic chemistry*, (Int. edn.). Holt Saunders, Japan.

Sharpe, A.G. (1992). *Inorganic chemistry*, (3rd edn.). Longman Scientific & Technical, Harlow, UK.

Shriver, D.F., Atkins, P.W., and Langford, C.H. (1994). *Inorganic chemistry*, (2nd edn.) Oxford University Press, Oxford, UK.

Smith, D.W. (1990). *Inorganic substances*. Cambridge University Press, Cambridge, UK.

The nature of chemistry: transition-metal chemistry 1 and 2, Units 12-16 (1976). J. Evans and I.C. Nuttall (Eds.). Open University Press, Milton Keynes, UK.

Webster, B. (1990). *Chemical bonding theory*. Blackwell Scientific Publications, Oxford, UK.

Wulfsberg, G. (1987). *Principles of descriptive inorganic chemistry*. Brooks/Cole Publishing Co., Monterey, USA.

Index